Workbook 7
Modeling and Simulation for Application Engineers

Dr. Medhat Kamel Bahr Khalil, Ph.D, CFPHS, CFPAI.
Director of Professional Education and Research Development,
Applied Technology Center, Milwaukee School of Engineering,
Milwaukee, WI, USA.

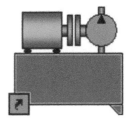

CompuDraulic LLC
www.CompuDraulic.com

1

CompuDraulic LLC

Workbook 7

Modeling and Simulation
for Application Engineers

ISBN: 978-0-9977634-4-7

Printed in the United States of America
First Published by June,2020
Revised by --

All rights reserved for CompuDraulic LLC.
3850 Scenic Way, Franksville, WI, 53126 USA.
www.compudraulic.com

Disclaimer

It is always advisable to review the relevant standards and the recommendations from the system manufacturer. However, the content of this book provides guidelines based on the author's experience.

Any portion of information presented in this book could be not applicable for some applications due to various reasons. Since errors can occur in circuits, tables, and text, the publisher assumes no liability for the safe and/or satisfactory operation of any system designed based on the information in this book.

The publisher does not endorse or recommend any brand name product by including such brand name products in this book. Conversely the publisher does not disapprove any brand name product by not including such brand name in this book. The publisher obtained data from catalogs, literatures, and material from hydraulic components and systems manufacturers based on their permissions. The publisher welcomes additional data from other sources for future editions.

Workbook 7
Modeling and Simulations for Application Engineers
Table of Contents

PREFACE

This Workbook is a complementary part to the textbook of the same title. This book is used as a workbook for students to take notes during the course delivery. It contains colored printout of the PowerPoint slides that are designed to present the course. Each chapter is followed by a number of review questions and assignments for homework.

Dr. Medhat Kamel Bahr Khalil

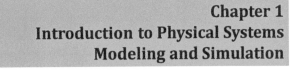

Chapter 1
Introduction to Physical Systems Modeling and Simulation

Objectives:

Modeling and simulation are essential tools in today's system design process. This chapter introduces the subject matter overviewing, the importance, historic background, and the challenges in physical systems modeling and simulation. The chapter also provides a brief overview of the common techniques used for mathematical modeling of physical systems in t-domain and s-domain. The chapter also presents the typical forcing functions used to simulate physical systems performance analysis under various load conditions or commands.

0

0

Brief Contents:

1.1- Importance of Physical Systems Modeling and Simulation
1.2- History of Physical Systems Modeling and Simulation
1.3- Product Development Cycle
1.4- Physical Systems Identification
1.5- Physical Systems Mathematical Modeling
1.6- Physical Systems Modeling in t-Domain using Differential Equations
1.7- Physical Systems Modeling in s-Domain using Laplace Transform
1.8- t-Doman versus s-Domain Physical Systems Modeling
1.9- Development of Simulation Models for Physical Systems
1.10- Physical Systems Performance Simulation
1.11- Physical Systems Performance Analysis
1.12- Challenges of Physical Systems Modeling and Simulation
1.13- Basic Elements of Physical Systems
1.14- Effort and Flow Variables of Physical Systems
1.15- Power Calculation for Physical Systems
1.16- Mathematical Representation of Inductive Elements
1.17- Mathematical Representation of Resistive Elements
1.18- Mathematical Representation of Capacitive Elements

1

1

1.1- Importance of Physical Systems Modeling and Simulation

- Design Optimization (cost, efficiency, reliability, etc.).
- Performance Investigation (pressure, temperature, etc.)
- System Prototyping (time & cost).
- Controller Prototyping (hardware-in-the-loop).

Fig. 1.1 – Various Industry Sectors

2

2

1.2- History of Physical Systems Modeling and Simulation

Previously:

Lack of computational tools → Mathematicians

→ Simplified mathematical solutions (assumptions + linearization)

→ Time and cost inefficient trial and error design strategy.

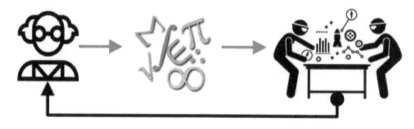

Fig. 1.2 – Physical Systems Design Methodology

3

3

Today:

Microprocessors (calculations even in Real-Time)

→ Modeling and Simulation

→ Virtual System.

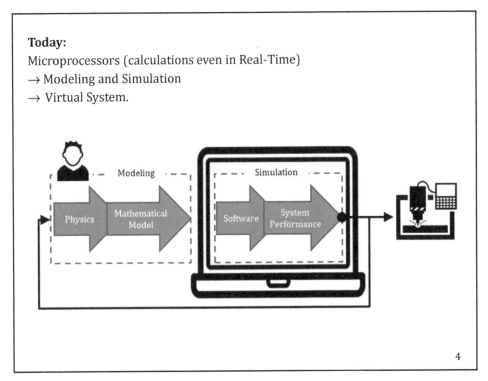

4

4

1.3- Product Development Cycle

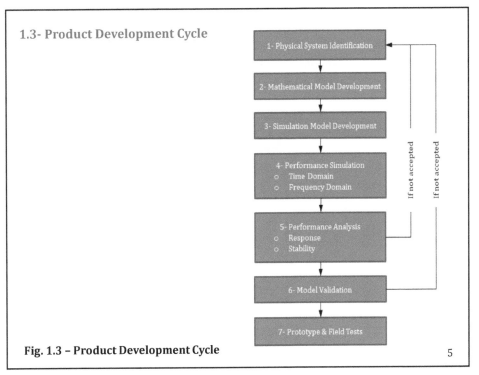

Fig. 1.3 – Product Development Cycle

5

5

1.4- Physical Systems Identification

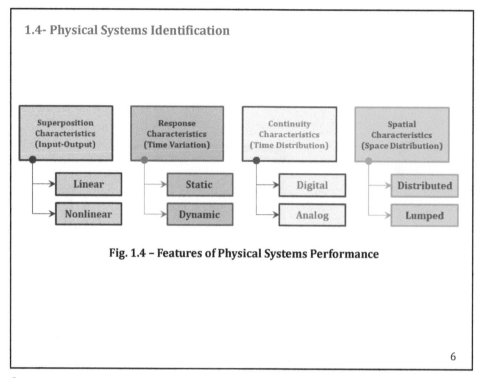

Fig. 1.4 – Features of Physical Systems Performance

6

6

1.4.1- Linear versus Nonlinear Systems

Linear $\rightarrow$ simple mathematical solutions $\rightarrow$ superposition

f(x1+x2) = f(x1) + f(x2) & f(kx) = kf(x)

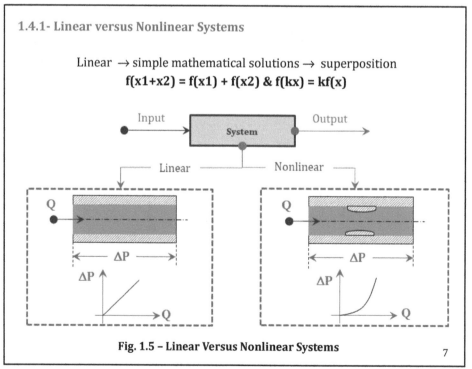

Fig. 1.5 – Linear Versus Nonlinear Systems

7

7

1.4.2- Static versus Dynamic Systems

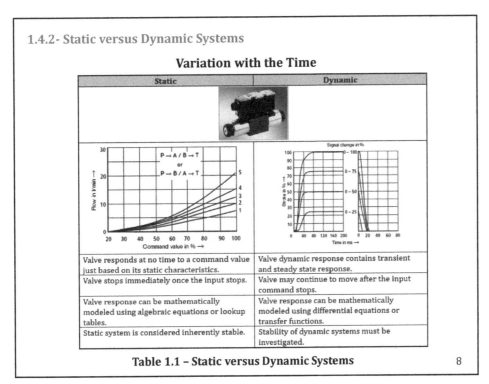

Table 1.1 – Static versus Dynamic Systems

8

Note:

- Be careful when connecting models for static and dynamic systems.
- It may result in errors when calculating the derivatives.
- Example: static model of a pump generate flow at no time to a dynamic valve model.

1.4.3- Digital versus Analog Systems

- **Digital (Binary) System:** Time/Value = Continuous/Binary
- **Analog (Continuous) System:** Time/Value = Continuous/Continuous
- **Discrete System:** Time/Value = Continuous/Discrete

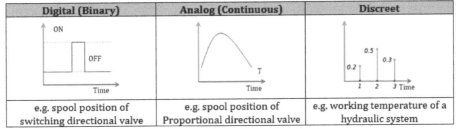

Digital (Binary)	Analog (Continuous)	Discreet
e.g. spool position of switching directional valve	e.g. spool position of Proportional directional valve	e.g. working temperature of a hydraulic system

Table 1.2 – Digital versus Analog Systems

9

1.4.4- Distributed versus Lumped Systems

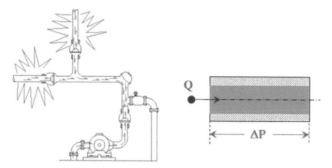

Fig. 1.6 – Distributed Versus Lumped Systems

1.4.5- Design Parameters and Assumptions

- Understand system kinematics and dynamics.
- Identify measurable design parameters
- Such as (mass, spring stiffness, dimensional parameters, etc.)
- Reasonably, and feasibly assume unmeasurable parameters
- Such as (friction coefficient, leakage coefficient, etc.)

10

10

1.5- Physical Systems Mathematical Modeling

System Identification → Mathematical Modeling

Physical Systems Mathematical Modeling in t-Domain:
- System represented by Differential Equations (DE).
- System performance is presented versus time.

Physical Systems Mathematical Modeling in s-Domain:
- System represented by Transfer Function using Laplace Transform.
- System performance is presented on a Complex Plane (Real and Imaginary).

Other Methods: Physical systems can be mathematically modeled using different techniques such as:
- **State-Space Representation:** Matrix Representation.
- **Bond-Graph Modeling:** Graphical and Object-Orients Representation.

11

11

1.6- Physical Systems Modeling in t-Domain using Differential Equations
1.6.1- Classifications of Differential Equations
1.6.1.1- Ordinary versus Partial Differential Equations

Example of an Ordinary DE: $\qquad \frac{dy}{dt} = 3\sin(t)$

Practical Example: Temperature at a pump outlet is function of time.

Example of a Partial DE: $\qquad \frac{\partial y}{\partial t} + \frac{\partial y}{\partial x} = at + bx$

Practical Example: Pressure varies with the time and the length of a transmission line.

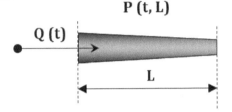

Fig. 1.7 – Potential Energy for Hydraulic Cylinders

12

12

1.6.1.2- Linear versus Nonlinear Differential Equations

Linear DE:
- All variables and their derivatives appear with first power.
- Mathematical solutions are well developed.

$$ax'' + bx' + cx = F(t)$$

Nonlinear DE:
- At least one of the variables or its derivative is raised to power > 1.
- Mathematical solutions are complex.

$$y' + \frac{1}{y} = 0 \;,\; y' + y^2 = 0 \;,\; y'' + \sin(y) = 0 \;,\; \text{and } yy' = 1$$

13

13

System Linearization **technique.**

$$y'' + 2y' + e^y = 5t$$

The operator $e^y = 1 + y + \frac{y^2}{2} + \frac{y^3}{6} + $ etc.

Then, by removing the nonlinear parts $\rightarrow$ $e^y = 1 + y$

Then the linearized system is mathematically modeled as follows:

$$y'' + 2y' + 1 + y = 5t$$

- Linearized model $\rightarrow$ minor details are not captured $\rightarrow$ less accurate in representing the real physical system.

- **Question:** Do system designers still need to develop linearized models?

- **Answer:** "not necessarily" because there are software tools that can solve the system model as is, no matter what the linearity state of the system is.
- However, linearized techniques may still be used to apply simple design methods and predict an approximate system response.

14

14

1.6.1.3- Homogeneous versus Non-Homogeneous Differential Equations
Non-Homogeneous DE: Ordinary DE contains *Forcing Function* **F(t).**

$$aX'' + bX' + cX = F(t)$$

Homogeneous DE: referred to as *Characteristic Equation.*

$$aX'' + bX' + cX = 0$$

1.6.1.4- First Order versus Second Order Differential Equations
First Order DE: $\qquad\qquad ax' + bx = F(t) \qquad\qquad 1.1$

Second Order DE: $\qquad\qquad ax'' + bx' + cx = F(t) \qquad\qquad 1.2$

Example of 2nd Order DE. A system contains the three basic physical elements (inductive, capacitive, and resistive).

Higher Order DE: $\qquad ax''' + bx'' + cx' + dx = F(t) \quad 1.3$

Example of Higher Order DE. A system in which a mass is accelerated by variable acceleration acting jerky.

15

15

1.6.2- Mathematical Solution in t–Domain for a Physical System Response

- Roots of the system indicates the system response and stability.
- Mathematical solution won't be discussed in this textbook.

Fig. 1.8 – Mathematical Solutions for Physical Systems

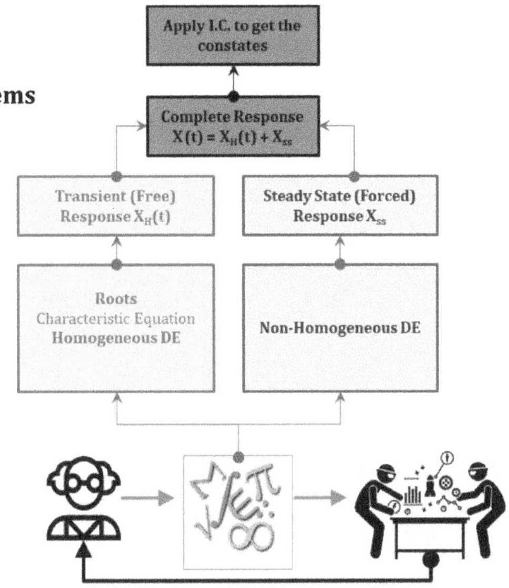

16

1.7- Physical Systems Modeling in s-Domain using Laplace Transform

Physical Systems Mathematical Modeling in s-Domain:
- System represented by Transfer Function using Laplace Transform.
- System performance is presented on a Complex Plane (**Real and Imaginary**).

Background:
The conceptual idea of Laplace Transform is to convert a differential equation into an algebraic equation that can be solved easily, then the solution can be reconverted using *Inverse Laplace Transform*.

Question: Do we still need to model systems using LT in form of TF?
Answer:
- This method became part of system design history.
- It is up to the system modeler based on his preference.
- This technique is helpful for large systems because subsystems can be capsulated in just one TF.
- Fortunately, simulation software can accept either or combination of t-domain and s-Domain models on the same layer of the system model.

17

1.7.1- Laplace Transform

DE in t-domain → equation in s-Domain
derivative → **s** & Integral → **"1/s"**.
$$\mathcal{L}\{F(t)\} = F(s) \qquad 1.4$$

1.7.2- Block Diagram Algebra

Example 1 (multiplying by a constant):

$Y(s) = K\,X(s)$

Example 2 (Derivative):

$Y(s) = s\,X(s)$

Example 3 (Integral):

$Y(s) = 1/s\,X(s)$

18

18

Example 4 (Addition and Subtraction):

$Y(s) = X_1(s) - X_2(s)$

Example 5 (Associative and Commutative Properties):

$Y(s) = G_1 G_2\,X(s)$

19

19

14

Example 6 (Distributive Property):

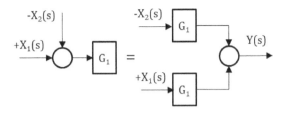

$$Y(s) = G_1X_1(s) - G_1X_2(s)$$

Example 7 (Blocks in Parallel):

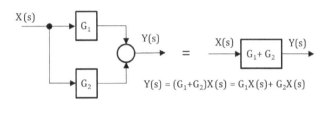

$$Y(s) = (G_1+G_2)X(s) = G_1X(s) + G_2X(s)$$

20

20

1.7.3- Transfer Function

block diagram $\rightarrow$ reduced to just one block $\rightarrow$ TF = output/input

Example 1: Positive Feedback Loop:

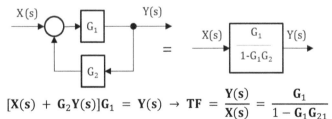

$$[X(s) + G_2Y(s)]G_1 = Y(s) \rightarrow TF = \frac{Y(s)}{X(s)} = \frac{G_1}{1 - G_1G_{21}}$$

Example 2: Negative Feedback Loop:

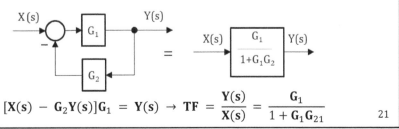

$$[X(s) - G_2Y(s)]G_1 = Y(s) \rightarrow TF = \frac{Y(s)}{X(s)} = \frac{G_1}{1 + G_1G_{21}}$$

21

21

1.8- t-Domain versus s-Domain Physical Systems Modeling

Question: Does the domain in which a physical system is modeled affect the system response"?

Note: Matlab-Simulink accept
t-domain and s-domain models on the same layer.

- **Use for Complex Systems:**
- **Optimize Design Parameters:**
- **System Performance Prediction:**

22

22

1.9- Development of Simulation Models for Physical Systems

System Identification → Mathematical Modeling
→ Performance Simulation

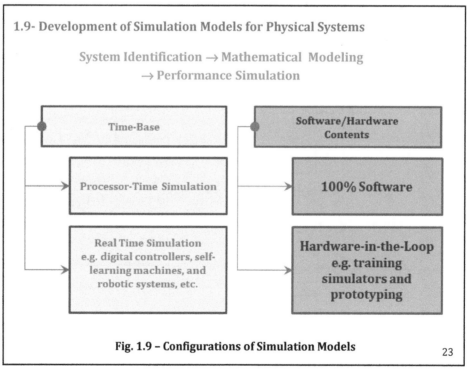

Fig. 1.9 – Configurations of Simulation Models

23

23

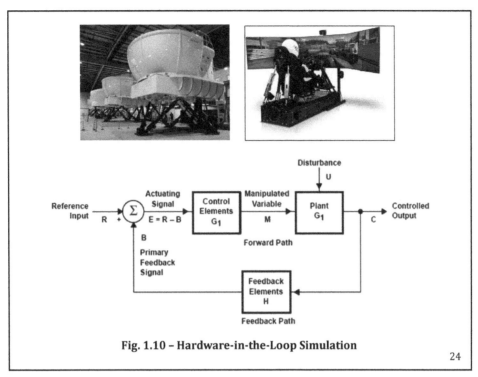

Fig. 1.10 – Hardware-in-the-Loop Simulation

24

24

1.10- Physical Systems Performance Simulation

System Identification → Mathematical Modeling → Performance Simulation

Name	Physical Interpretations	F(t)	F(s)	Graphical Representation
Impulse	Represents shock loads or commands that occur in a very short time, such as pressure spikes or opening and closing an on/off valve.	$\delta(t)$	k	$F(t)$ ↑ → t
Step	Represents sudden loading or commands to a constant value, such as requesting a variable pump to go from zero to stay at 75% displacement in a very short time.	k	k/s	$F(t)$ ↑ → t
Ramp	Represents gradual loading or commands with constant rate, such as opening of a proportional DCV over a relatively long time e.g. 3 seconds. Usually used to develop static characteristic of a system.	kt	k/s^2	$F(t)$ ↑ → t
Harmonic	Physical system must be tested for periodic loads. Pure harmonic signal is a result of Fourier Analysis of periodic load. If a system passes the test under harmonic signal, it should pass the test under periodic load.	$\sin(\omega t)$ & $\cos(\omega t)$	$\omega/(s^2 + \omega^2)$ & $s/(s^2 + \omega^2)$	$F(t)$ ↑ → t

Table 1.3 – Typical Forcing Functions

25

25

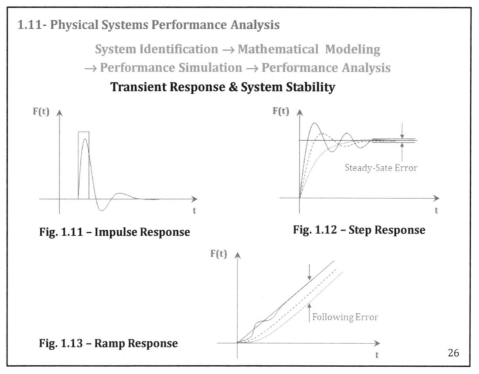

1.11- Physical Systems Performance Analysis

System Identification → Mathematical Modeling
→ Performance Simulation → Performance Analysis

Transient Response & System Stability

Fig. 1.11 – Impulse Response

Fig. 1.12 – Step Response

Fig. 1.13 – Ramp Response

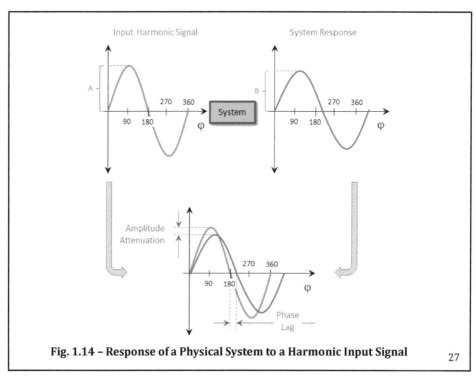

Fig. 1.14 – Response of a Physical System to a Harmonic Input Signal

$$AR[dB] = 20 \log \left[\frac{B}{A}\right] \qquad 1.5$$

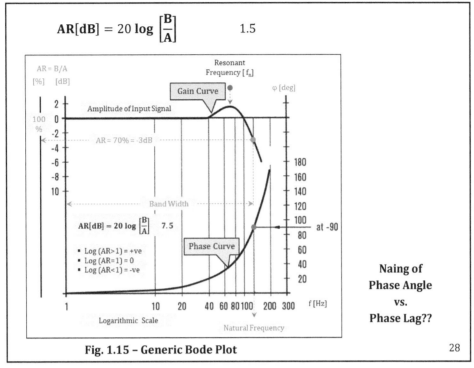

Fig. 1.15 – Generic Bode Plot

28

28

1.12- Challenges of Physical Systems Modeling and Simulation
1.12.1- Challenges in Physical System Identification

- **Features of the Physical System:**
  o Clearly understand the features (linear or nonlinear, static or dynamic, digital or analog, and distributed or lumped).

- **Component Design Parameters:**
  o Access to classified design parameters.

- **Assumptions:**
  o Unmeasurable parameters.
  o lack of information.
  o Feasible and reasonable assumptions.

29

29

1.12.2- Challenges in Mathematical Model Development

- **Level of Model Complexity:**
  - It is a case-by-case, tailoring process.
  - A model shouldn't be tight (simplified) → missing details
  - A model shouldn't be large (exaugurated) → giving unnecessary details wasting valuable resources of time and cost.

- **Mathematical Modeling Technique:**
  - (DE) or (TF)? it depends on the know-how of the system modeler.

- **Solvable Model:**
  - Developing a solvable mathematical model.
  - No Algebraic Loops OR Overflow (division by zero)

30

30

1.12.3- Challenges in Simulation Model Development
- **Input/output Parameters:**
- **Models for Static vs. Dynamics:**
- **Programming Language:**
- **Artwork:**

1.12.4- Challenges in Performance Simulation
- **Sampling Rate:**
- Small → missing data, Large → increased computational time and effort.
- Fixed or Variable?
- **Input Signals:** impulse, step, ramp, harmonic?

1.12.5- Challenges in Performance Analysis
- **Model Results:**
  - Does the results are realistic?
  - A modeler shouldn't be slave to the math.
- **Model Debugging:**
- A modeler should be aware of the debugging tools.

31

31

1.13- Basic Elements of Physical Systems

Elements→ System↓	Inductive	Resistive	Capacitive
Mechanical Translational	Translational Mass (m)	k_f Linear Damper Damping Coefficient (k_f)	C Linear Spring Capacitance (C) = 1/Spring Stiffness (k_s)
Mechanical Rotational	J Rotational Mass Mass Moment of Inertia (J)	k_f Torsional Damper Damping Coefficient (k_f)	C Torsion Spring Capacitance (C) = 1/Spring Stiffness (k_s)

Table 1.4 – Basic Elements of Physical Systems   32

32

Elements→ System↓	Inductive	Resistive	Capacitive
Hydraulic	I Fluid Line Inductance (I)	R Fluid Line Resistance (R)	C Fluid Line Capacitance Capacitance (C) = V_0/Bulk Modulus (β)
Electrical	L Electric Inductor Inductance (L)	R Electric Resistor Resistance (R)	C Electrical Capacitor Capacitance (C)
Thermal	Not Applicable	Heat Resistor Resistance (R)	Heat Sink Capacitance (C)

33

33

1.14- Effort and Flow Variables of Physical Systems

Physical systems behave similarly

If $|$Effort Variable$| > 0 \rightarrow |$Flow Variable$| > 0$

If $|$Effort Variable$| = 0 \rightarrow |$Flow Variable$| = 0$ OR $=$ Constant

Charge Variable $= \int($Effort Variable$)$

❖ **Mechanical Translational System:**

- Effort Variable = *Resultant Force* **F** & Flow Variable = *Linear Speed* **v**.
- If $|\Sigma F| > 0 \rightarrow |v| > 0$.
- i.e. Mass **m** is either accelerating or decelerating depends on sign of **ΣF**.
- If $|\Sigma F| = 0 \rightarrow v = 0$ OR $|v| =$ Constant.
- i.e. Mass **m** is either standstill or translating with constant **v**.
- Charge Variable is *Linear Displacement* $x = \int v\,dt$.

Variables→ Systems↓	Effort Variable	Flow Variable	Charge Variable	
Mechanical Translational	Resultant Force F	Linear Velocity v	Linear Displacement x	

❖ **Mechanical Rotational System:**

- Effort Variable = *Resultant Torque* **T** & Flow Variable = *Angular Speed* **ω**.
- If $|\Sigma T| > 0 \rightarrow |\omega| > 0$.
- i.e. Moment of Inertia **J** is either accelerating or decelerating depends on the sign of **ΣT**.
- If $|\Sigma T| = 0 \rightarrow \omega = 0$ OR $|\omega| =$ Constant.
- i.e. **J** is either standstill or rotating with constant **ω**.
- Charge Variable is *Angular Displacement* $\theta = \int \omega\,dt$.

Variables→ Systems↓	Effort Variable	Flow Variable	Charge Variable	
Mechanical Rotational	Resultant Torque T	Angular Velocity ω	Angular Displacement θ	

Table 1.5 - Effort and Flow Variables for Physical Systems

❖ **Hydraulic System:**
- Effort Variable = *Differential Pressure* **ΔP.**
- Flow Variable is *Fluid Flow* **Q**.
- If $|\Delta P| > 0 \rightarrow$ Fluid Flow **Q** moves in a direction depends on the sign of **ΔP.**
- Charge Variable is *Fluid Volume* $\Delta V = \int Qdt$.

Variables→ Systems↓	Effort Variable	Flow Variable	Charge Variable	
Hydraulic	Differential Pressure **ΔP**	Fluid Flow **Q**	Fluid Volume **ΔV**	

36

❖ Electrical System:
- Effort Variable = *Voltage Difference* **Δv.**
- Flow Variable is *Electrical Current* **i**.
- If $|\Delta v| > 0 \rightarrow$ Electrical Current **i** moves in a direction depends on the sign of **Δv**.
- Charge Variable is *Electrical Charge* $Q = \int idt$.

Variables→ Systems↓	Effort Variable	Flow Variable	Charge Variable	
Electrical	Voltage Difference **Δv**	Electrical Current **i**	Electrical Charge **Q**	

37

❖ Thermal System:
- Effort Variable = *Temperature Difference* ΔT
- Flow Variable = *Heat Flow* **q**.
- If $|\Delta T| > 0 \rightarrow$ Heat Flow **q** moves in a direction depends on the sign of ΔT.
- Charge Variable is *Heat* $Q = \int q \, dt$.

Variables→ Systems↓	Effort Variable	Flow Variable	Charge Variable	
Thermal	Temperature Difference ΔT	Heat Flow q	Heat Q	

38

38

1.15- Power Calculation for Physical Systems

$$\text{Power} = \text{Effort Variable} \times \text{Flow Variable} \quad ----- 1.5$$

❖ For a Mechanical Translational System:
$$\text{Power}, H = F \times v \qquad\qquad 1.5A$$

❖ For a Mechanical Rotational System:
$$\text{Power}, H = T \times \omega \qquad\qquad 1.5B$$

❖ For a Hydraulic System:
$$\text{Power}, H = \Delta P \times Q \qquad\qquad 1.5C$$

❖ For an Electrical System:
$$\text{Power}, H = \Delta v \times i \qquad\qquad 1.5D$$

39

39

Note: As shown in Fig. 1.16, same equation (Eq. 1.5C) is used to quantify the hydraulic power whether the power is gained by a pump, consumed to do useful work by a hydraulic actuator, or wasted as heat in an element.

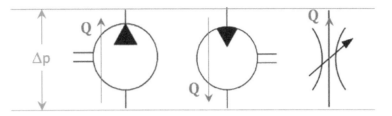

Fig. 1.16 – Power Calculation for Hydraulic Components

40

40

1.16- Mathematical Representation of Inductive Elements

Effort Variable = Inductance × Rate of change of Flow Variable − 1.6

❖ For a Mechanical Translational System:

$$\mathbf{F} = \mathbf{m} \times \frac{\mathbf{dv}}{\mathbf{dt}} \qquad \qquad \mathbf{1.6A}$$

❖ For a Mechanical Rotational System:

$$\mathbf{T} = \mathbf{J} \times \frac{\mathbf{d\omega}}{\mathbf{dt}} = \mathbf{J} \times \mathbf{\alpha} \qquad \qquad \mathbf{1.6B}$$

❖ For an Electrical System:

$$\mathbf{\Delta v} = \mathbf{L} \times \frac{\mathbf{di}}{\mathbf{dt}} \qquad \qquad \mathbf{1.6C}$$

Inductance **L** of an electrical inductor is a given value depending on the design of the inductor.

❖ For a Hydraulic System:

$$\mathbf{\Delta P} = \mathbf{I} \times \frac{\mathbf{dQ}}{\mathbf{dt}} \qquad \qquad \mathbf{1.6D}$$

41

41

Inductance I of a hydraulic transmission line

$$\Delta P = \frac{\text{Fluid Inertia}}{\text{Line Area (A)}} = \frac{\text{Fluid Mass (m)}}{\text{Line Area (A)}} \times \frac{d\,[\text{Fluid Velocity (v)}]}{dt}$$

Since Fluid Flow Q = v x A, and
Fluid Mass = Fluid Density ρ x Fluid Volume V, then

$$\Delta P = \frac{m}{A} \times \frac{1}{A} \times \frac{dQ}{dt} = \frac{\rho \times V}{A} \times \frac{1}{A} \times \frac{dQ}{dt} = \frac{\rho \times A \times L}{A} \times \frac{1}{A} \times \frac{dQ}{dt} = \frac{\rho L}{A}\frac{dQ}{dt} = I\frac{dQ}{dt}$$

$$\text{Then } I = \frac{\rho L}{A} \qquad\qquad 1.7$$

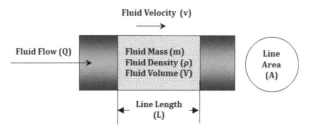

Fig. 1.17 – Inductance of a Hydraulic Transmission Line 42

Kinetic Energy (associated with inductive elements):

- Kinetic Energy is the energy an object possesses due to its motion.
- It is defined as the work needed to accelerate a body of a given mass from one velocity to other velocity.

$$\textbf{Kinetic Energy} = \tfrac{1}{2}\,\textbf{Inductance} \times (\textbf{Flow Variable})^2 \quad ----- 1.8$$

❖ For a Mechanical Translational System:

$$E_K = \frac{1}{2}mv^2 \qquad 1.8A$$

❖ For a Mechanical Rotational System:

$$E_K = \frac{1}{2}J\omega^2 \qquad 1.8B$$

❖ For an Electrical System:

$$E_K = \frac{1}{2}Li^2 \qquad 1.8C$$

❖ For a Hydraulic System:

$$E_K = \frac{1}{2}mv^2 = \frac{1}{2}(\rho V)\left(\frac{Q}{A}\right)^2 = \frac{1}{2}(\rho AL)\left(\frac{Q}{A}\right)^2 = \frac{1}{2}\left(\frac{\rho L}{A}\right)Q^2 = \frac{1}{2}IQ^2 \qquad 1.8D$$

43

1.17- Mathematical Representation of Resistive Elements

$$\text{Effort Variable} \;=\; \text{Resistivity Factor} \;\times\; \text{Flow Variable} \;---- 1.9$$

❖ For a Mechanical Translational System (Hook's Law):

$$F = k_f v \qquad\qquad 1.9A$$

❖ For a Mechanical Translational System:

$$T = k_f \omega \qquad\qquad 1.9B$$

❖ For an Electrical System:

$$\Delta v = Ri \qquad\qquad 1.9C$$

❖ For a Hydraulic Transmission Lines:

$$\Delta P = RQ \qquad\qquad 1.9D$$

❖ For a Hydraulic Valves (Nonlinear):

$$\Delta P = RQ^2 \qquad\qquad 1.9E$$

44

44

Dissipated (Wasted) Energy (associated with resistive elements):

- Energy losses are relative because energy can't vanish.
- It is the energy that is converted into a non-useful form.
- Energy losses in hydraulic components are converted into heat.
- In an electrical heater, any energy that does not converted into heat is considered wasted energy!

$$\textbf{Wasted Energy} = \textbf{Wasted Power} \times \textbf{Time}$$

Equation 1.5 →

$$\textbf{Wasted Energy} = \textbf{Effort Variable} \times \textbf{Flow Variable} \times \textbf{Time}$$

Then

$$\text{Wasted Energy} = \text{Effort Variable} \times \text{Charge Variable} \;---- 1.10$$

45

45

❖ For a Mechanical Translational System (Hook's Law):

$$E_W = Fx = k_f v x \qquad\qquad 1.10A$$

❖ For a Mechanical Rotational System:

$$E_W = T\theta = k_f \omega \theta \qquad\qquad 1.10B$$

❖ For an Electrical System:

$$E_W = \Delta v Q = RiQ \qquad\qquad 1.10C$$

❖ For a Hydraulic Transmission Lines:

$$E_W = \Delta P \Delta V = RQ\Delta V \qquad\qquad 1.10D$$

❖ For a Hydraulic Valves (Nonlinear):

$$E_W = \Delta P \Delta V = RQ^2\Delta V \qquad\qquad 1.10E$$

46

46

1.18- Mathematical Representation of Capacitive Elements

$$\text{Effort Variable} = \frac{1}{\text{Capacitnace}} \int (\text{Flow Variable})\, dt = \frac{\text{Charge Variable}}{\text{Capacitnace}} \quad --1.11$$

❖ For a Mechanical Translational System (Hook's Law):

$$F = \frac{1}{C} \int v\, dt = k_s x \qquad\qquad 1.11A$$

Where k_s is the spring stiffness and x is the linear deformation of a linear spring

❖ For a Mechanical Rotational System:

$$T = \frac{1}{C} \int \omega\, dt = k_s \theta \qquad\qquad 1.11B$$

Where k_s is the spring stiffness and θ is the angular deformation of a torsion spring.

For an Electrical System:

$$\Delta v = \frac{1}{C} \int i\, dt = \frac{\text{Erlecrtic Charge (Q)}}{C} \qquad\qquad 1.11C$$

For a Hydraulic System:

$$\Delta P = \frac{1}{C} \int Q\, dt = \frac{\text{Fluid Bulck Modulus }(\beta)}{\text{Container Volume }(V_0)} \int Q\, dt = \frac{\beta \Delta V}{V_0} \qquad\qquad 1.11D$$

This equation can be used to investigate pressure spikes

47

47

Potential (stored) Energy (associated with capacitive elements):

- Potential Energy of an object is the energy stored in the object because of its altitude relative to other objects, elastic deformation, fluid compressibility, or electric charge.
- It is defined as the work needed to make such a change in the object.

$$\text{Potential Energy} = \frac{1}{2}\text{Effort Variable} \times \text{Charge Variable}$$

$$= \frac{(\text{Charge Variable})^2}{2 \times \text{Capacitnace}} \qquad 1.12$$

48

48

❖ For a Mechanical Translational System:

$$E_P = \frac{1}{2}Fx = \frac{x}{2}\left[\frac{1}{C}\int v\, dt\right] = \frac{x^2}{2C} = \frac{1}{2}k_s x^2 \qquad 1.12A$$

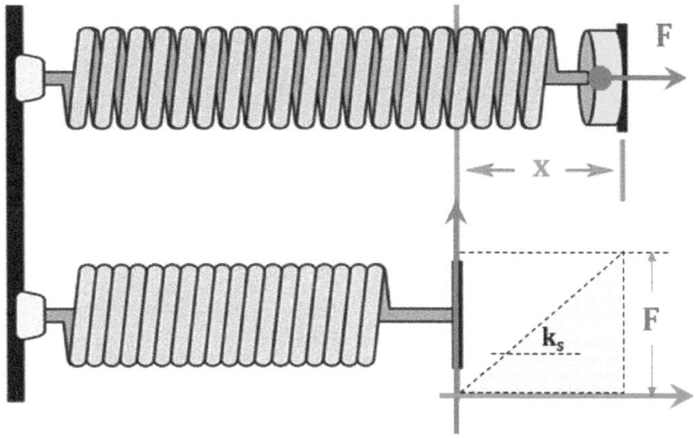

Fig. 1.18 – Potential Energy for Linear Spring

49

49

❖ For a Mechanical Rotational System:

$$E_P = \frac{1}{2}T\theta = \frac{\theta}{2}\left[\frac{1}{C}\int \omega \, dt\right] = \frac{\theta^2}{2C} = \frac{1}{2}k_s\theta^2 \qquad 1.12B$$

❖ For an Electrical System:

$$E_P = \frac{1}{2}\Delta vQ = \frac{Q}{2}\left[\frac{1}{C}\int i \, dt\right] = \frac{1}{2}\frac{1}{C}Q^2 \qquad 1.12C$$

50

50

❖ For a Hydraulic System:

$$E_P = \frac{1}{2}\Delta p\Delta V = \frac{\Delta V}{2}\left[\frac{1}{C}\int Q \, dt\right] = \frac{1}{2}\frac{1}{C}(\Delta V)^2 = \frac{1}{2}\frac{\beta}{V_0}(\Delta V)^2 \qquad 1.12D.1$$

$$E_P = \frac{1}{2}\Delta p\Delta V = \frac{1}{2}\Delta pAx = \frac{1}{2}Fx \qquad 1.12D.2$$

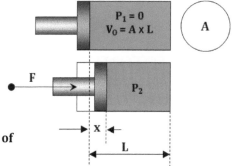

Fig. 1.19 – Potential Energy of Hydraulic Fluids

51

51

Chapter 1 Reviews

1. In a product cycle development, which of the following order of steps is considered correct?
    A. Physical system identification, mathematical modeling, simulation modeling, performance simulation, prototyping and field tests, performance analysis, model validation.
    B. Mathematical modeling, simulation modeling, performance simulation, performance analysis, model validation, prototyping and field tests, Physical system identification.
    C. Performance analysis, physical system identification, mathematical modeling, simulation modeling, performance simulation, model validation, prototyping and field tests.
    D. Physical system identification, mathematical modeling, simulation modeling, performance simulation, performance analysis, model validation, prototyping and field tests.

2. Which of the following statements is considered **FALSE**?
    A. Nonlinear system complies with the superposition characteristics
    B. Dynamic system contains transient and steady state response.
    C. Value of an analog system is continuous over the time.
    D. Value of a distributed system is distributed over the space.

3. Physical system modeling in s-Domain is more suitable for?
    A. A component level model to investigate effect of design parameters on component performance.
    B. A simple system that contains few components.
    C. A large system that contains multiple components.
    D. Only nonlinear systems.

4. Physical system modeling in t-Domain is more suitable for?
    A. A component level model to investigate effect of design parameters on component performance.
    B. A simple system that contains few components.
    C. A large system that contains multiple components.
    D. Only nonlinear systems.

5. Which of the following input signals are required to develop static characteristics of a system?
    A. Impulse.
    B. Ramp.
    C. Step.

D. Harmonic.

6. Which of the following input signals are required to simulate response of a system to a sudden command?
   A. Impulse.
   B. Ramp.
   C. Step.
   D. Harmonic.

7. Bandwidth of a system is the frequency at which the amplitude attenuated to?
   A. 40% of the exciting amplitudes.
   B. 50% of the exciting amplitudes.
   C. 60% of the exciting amplitudes.
   D. 70% of the exciting amplitudes.

8. A capacitive element in hydraulic systems is?
   A. A flow control valve.
   B. An accumulator.
   C. A pressure control valve.
   D. A reservoir.

9. Effort variable in hydraulic systems is?
   A. Flow rate
   B. Fluid Viscosity.
   C. Differential pressure across a component.
   D. Inlet pressure of a component.

10. Effort variable in hydraulic systems is?
    A. Flow rate
    B. Fluid Viscosity.
    C. Differential pressure across a component.
    D. Inlet pressure of a component.

Chapter 1 Assignment

Student Name: -- Student ID: ------------------

Date: -- Score: ------------------------

Assignment: Based on what you have learned about physical system identification, describe how are the systems classified based on input-output, time variation, time distribution, and space distribution.

Chapter 2
Modeling and Simulation of
First-Order Dynamic Systems

Objectives:

In this chapter, methods and theories presented in Chapter 1 are applied to First-Order dynamic systems. The chapter presents mathematical modeling for first-order systems in t-domain and s-domain. This chapter also presents the response of first-order systems to the typical forcing functions including, step, ramp, and harmonic inputs. The chapter discusses the measured characteristics of first-order step response and how to develop the transfer function of the system based on existing dynamic characteristics.

0

0

Brief Contents:

2.1- First-Order Physical System Identification

2.2- First-Order System Mathematical Modeling

2.3- Simulation Model Development of First-Order Systems

2.4- Performance Simulation of First-Order Systems

2.5- Step Response Analysis of First-Order Systems

2.6- First-Order System Identification Based on Step Response

2.7- Frequency Response Analysis of First-Order Systems

2.8- First-Order System Identification Based on Frequency Response

1

1

2.1- First-Order Physical System Identification

First-Order system contains two elements

(inductive & resistive) OR (capacitive & resistive).

 Why a system should contain a resistive element?

 Anim045

2

2

Example 1:
- Solve for the displacement **x.**
- Spring stiffness k_s & friction coefficient k_f must be known.

Example 2:
- Solve for the velocity **v.**
- Mass **m** & and the friction coefficient k_f must be known.
  k_f is not an easy to measure $\rightarrow$ **reasonably assumed and optimized**

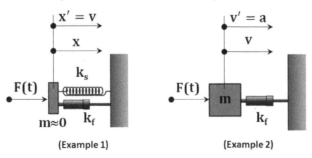

Fig. 2.1 – Examples of Mechanical Translational First-Order Systems

3

3

2.2- First-Order System Mathematical Modeling
2.2.1- First-Order System Mathematical Model in t-domain

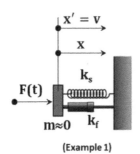

(Example 1)

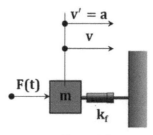

(Example 2)

$$k_f v + k_s x = F(t) \quad 2.1$$

steady state is when the rate of change of **x** with the time, i.e. the velocity **v**, equals zero.

$$x_{ss} = F(t)/k_s$$

$$ma + k_f v = F(t) \quad 2.2$$

steady state is when the rate of change of **v** with the time, i.e. the acceleration **a**, equals zero.

$$v_{ss} = F(t)/k_f$$

4

4

2.2.2- First-Order System Mathematical Model in s-domain

Applying L .Transform to D.E. 1.1→ $\quad (as + b)\, x(s) = F(s) \quad\quad 2.3$

τ (Time Constant) = a/b $\quad\quad (\tau s + 1)\, x(s) = \dfrac{F(s)}{b} \quad\quad 2.4$

$$TF = \frac{\text{Output}}{\text{Input}} = \frac{x(s)}{F(s)} = \frac{1/b}{\tau s + 1} \quad\quad 2.5$$

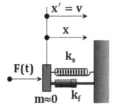

(Example 1)

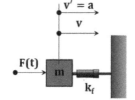

(Example 2)

$$TF = \frac{x(s)}{F(s)} = \frac{1/k_s}{\tau s + 1} \quad\quad 2.6$$

$$TF = \frac{v(s)}{F(s)} = \frac{1/k_f}{\tau s + 1} \quad\quad 2.7$$

$$\tau = k_f/k_s$$

$$\tau = m/k_f$$

5

5

To calculate the steady state value in s-domain, the Final Value Theorem applies as follows:

$$x_{ss} = \lim_{s \to 0} s.x(s) = \lim_{s \to 0} \frac{s.F(s)/b}{\tau s + 1} \qquad 2.8$$

$$\text{Step Input} \to F(s) = F/s \to \quad x_{ss} = \lim_{s \to 0} \frac{s.\frac{F}{s}/b}{\tau s + 1} = \frac{F}{b}$$

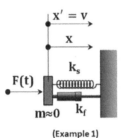

(Example 1)

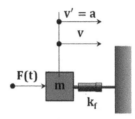

(Example 2)

$$x_{ss} = F(t)/k_s \qquad\qquad v_{ss} = F(t)/k_f$$

6

6

2.2.3- First-Order System Normalized Transfer Function

$$TF = \frac{Output}{Input} = \frac{1/b}{\tau s + 1} \quad \text{When } b = 1 \to \quad NTF = = \frac{1}{\tau s + 1} \qquad 2.9$$

Normalized Transfer Function (NTF)
Magnitude of the steady state variable = Magnitude of the forcing function F.

NTF is equivalent to a numerical value of 1

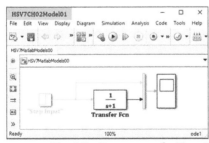

Fig. 2.2 – Response of a First-Order System Represented by TF

7

7

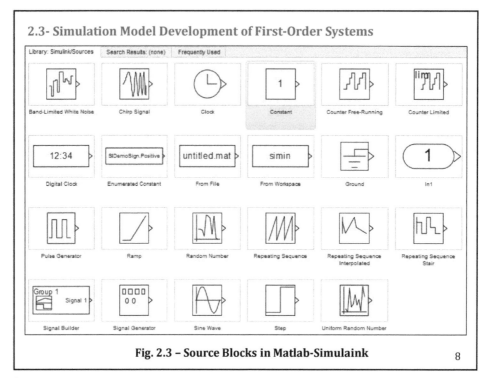

Fig. 2.3 – Source Blocks in Matlab-Simulaink

8

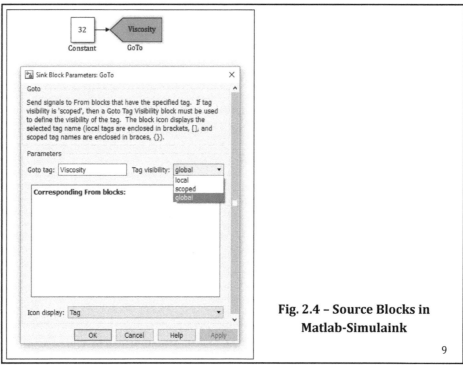

Fig. 2.4 – Source Blocks in Matlab-Simulaink

9

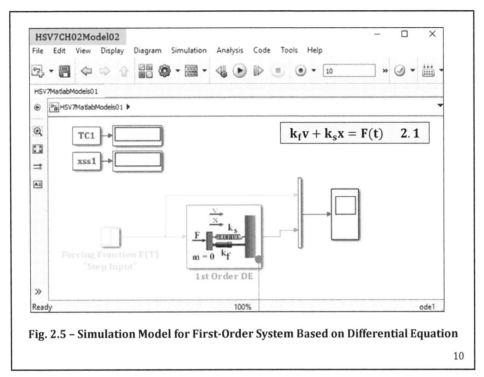

Fig. 2.5 – Simulation Model for First-Order System Based on Differential Equation

10

10

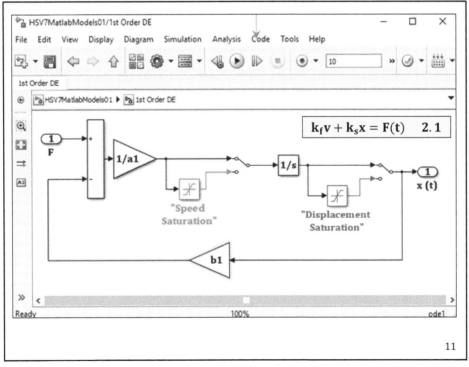

11

11

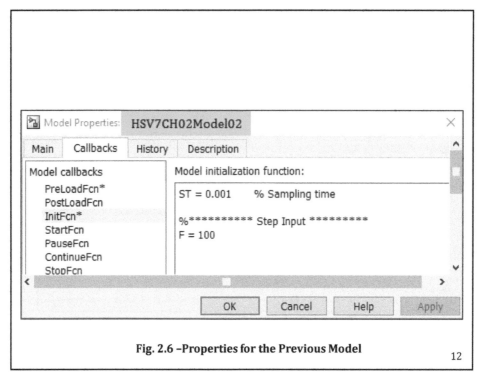

Fig. 2.6 –Properties for the Previous Model

12

12

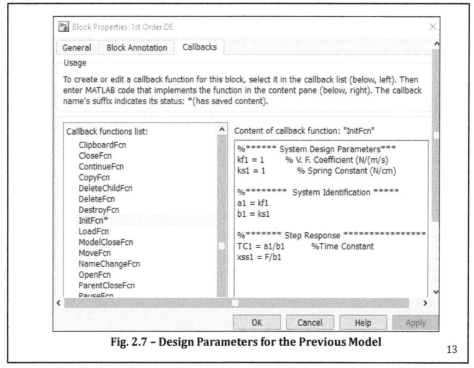

Fig. 2.7 – Design Parameters for the Previous Model

13

13

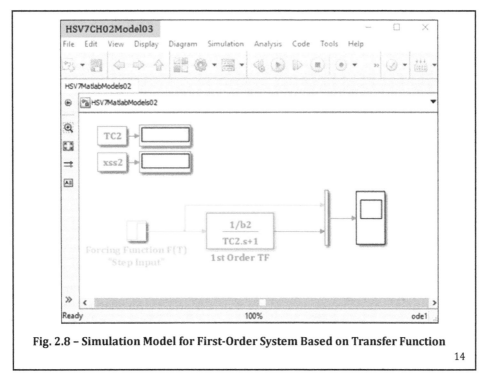

Fig. 2.8 – Simulation Model for First-Order System Based on Transfer Function

14

14

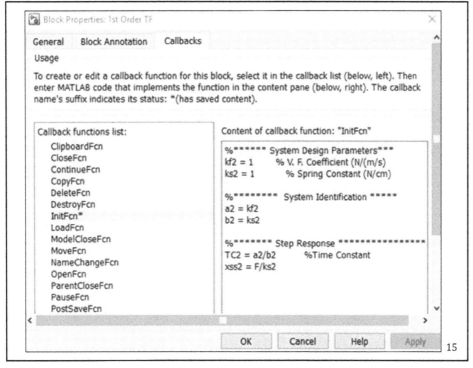

15

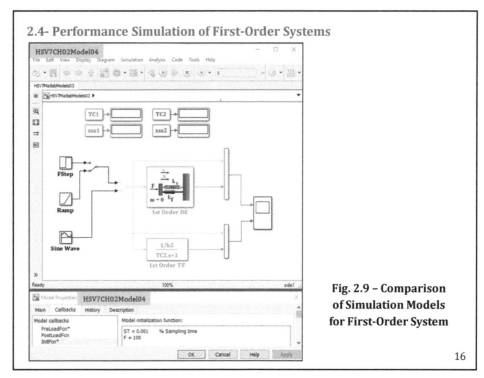

Fig. 2.9 – Comparison
of Simulation Models
for First-Order System

16

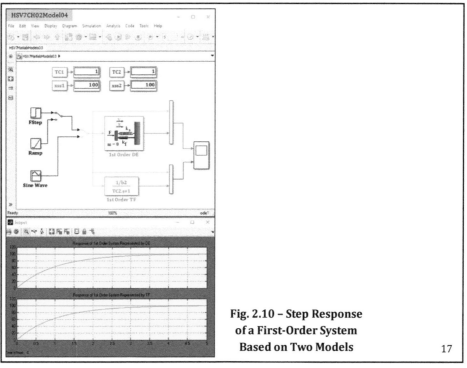

Fig. 2.10 – Step Response
of a First-Order System
Based on Two Models

17

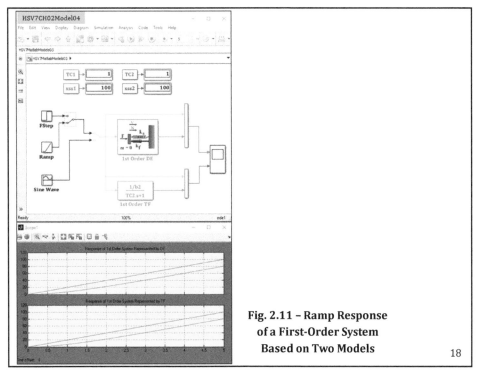

**Fig. 2.11 – Ramp Response
of a First-Order System
Based on Two Models**

18

18

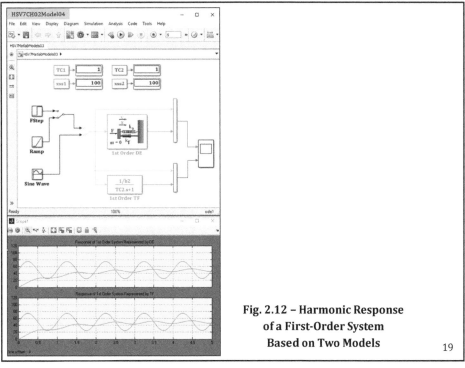

**Fig. 2.12 – Harmonic Response
of a First-Order System
Based on Two Models**

19

19

2.5- Step Response Analysis of First-Order Systems
2.5.1- Identification of First-Order Systems Based on Step Response

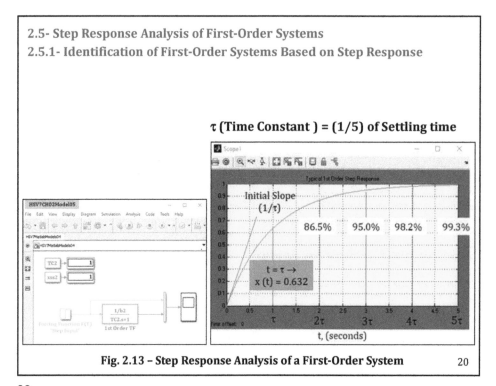

Fig. 2.13 – Step Response Analysis of a First-Order System

20

20

2.5.2- Effect of Design Parameters on First-Order System Step Response
2.5.2.1-Effect of Time Constant

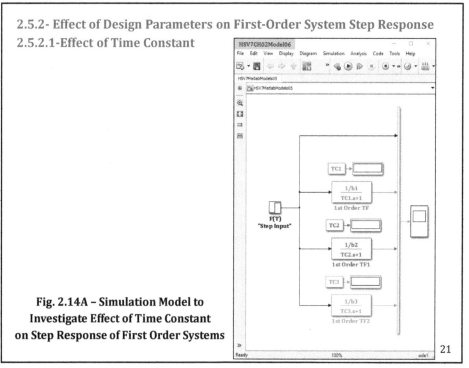

**Fig. 2.14A – Simulation Model to
Investigate Effect of Time Constant
on Step Response of First Order Systems**

21

21

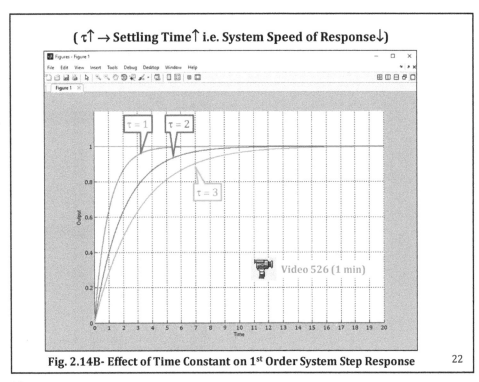

Fig. 2.14B- Effect of Time Constant on 1ˢᵗ Order System Step Response 22

22

2.5.2.2-Effect of Friction Coefficient

- **In Example 1:** $\tau = k_f/k_s$ and $x_{SS} = F/k_s$ (Hooke's law)
- $k_f \uparrow \rightarrow \tau \uparrow \rightarrow$ system speed of response $\downarrow$, i.e. slower response.
- $k_s \uparrow \rightarrow \tau \downarrow \rightarrow$ system speed of response $\uparrow$, i.e. system reaches x_{SS} faster.

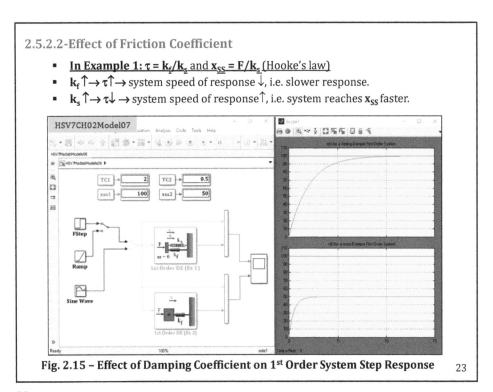

Fig. 2.15 – Effect of Damping Coefficient on 1ˢᵗ Order System Step Response 23

23

- **In Example 2:** $\tau = m/k_f$ and $v_{SS} = F/k_f$
- $m \uparrow \rightarrow \tau \uparrow \rightarrow$ system speed of response $\downarrow$, i.e slower response.
- $k_f \uparrow \rightarrow \tau \downarrow \rightarrow$ system speed of response $\uparrow$, i.e. system reaches to v_{SS} faster.

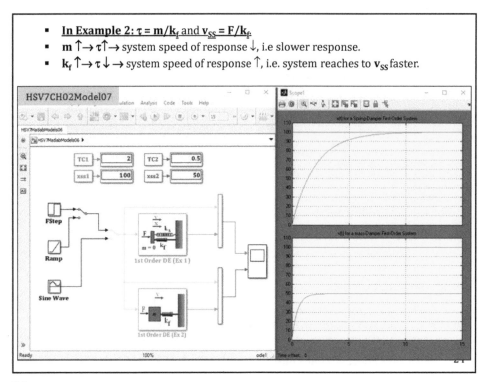

24

2.6- First-Order System Modeling Based on Step Response

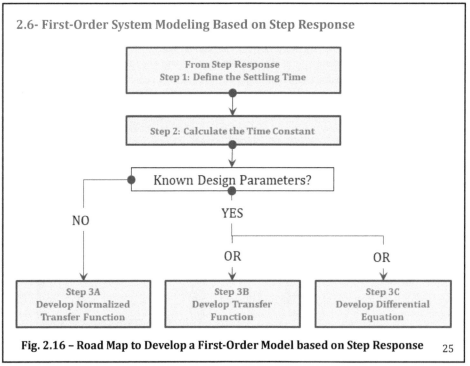

Fig. 2.16 – Road Map to Develop a First-Order Model based on Step Response 25

25

Step 1: Define the Settling Time:
- Settling Time is reported graphically or numerically.
- If not, it can also be found experimentally.

Step 2: Calculate the Time Constant:

Time constant τ equals one fifth of the settling time.

Step 3A: Develop the NTF
(in cases where physical design parameters are not known):

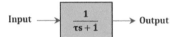

Fig. 2.17 – First-Order Model based on
Normalized Transfer Function

Step 3B: Develop the TF
(in cases where physical design parameters are known):

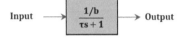

Fig. 2.18 – First-Order Model based on
Transfer Function

26

26

Step 3C: Developing DE (Alternative to Step 3B)
(in cases where physical design parameters are known):

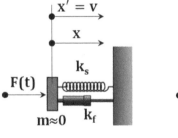

(Example 1)

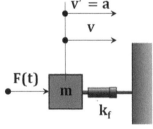

(Example 2)

$$k_f v + k_s x = F(t)$$

a = k_f "not measurable"
b = k_s "measurable"

$$\tau = a/b = k_f / k_s$$
Then $k_f = k_s \times \tau$

$$m a + k_f v = F(t)$$

a = m "measurable"
b = k_f "not measurable"

$$\tau = a/b = m / k_f$$
Then $k_f = m/\tau$

Fig. 2.19 – First-Order Model based on Differential Equation

27

27

Case Study:

Step 1: Define the Settling Time:

- F= 50 N. & m = 2000 kg. & Reported Settling Time =10 s.

Step 2: Calculate the Time Constant: $\tau = 10/5 = 2$ sec.

Step 3B: Construct the Model based on TF + Measurable Design

- Since $\tau = m/k_f$, then $k_f = m/\tau = 2000/2 = 1000$ (kg/s).

$$TF = \frac{Output}{Input} = \frac{v(s)}{F(s)} = \frac{1/k_f}{\tau s + 1} = \frac{1/1000}{2s + 1} \qquad 2.10$$

OR Step 3C:

Construct the Model based on Developing DE:

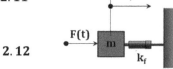

$$m[kg] \times a\left[\frac{m}{s^2}\right] + k_f \left[\frac{kg}{s}\right] \times v \left[\frac{m}{s}\right] = F(t) \ [N] \quad 2.11$$

$$2000 \ a\left[\frac{m}{s^2}\right] + 1000 \ v \left[\frac{m}{s}\right] = F(t) \ [N] \qquad 2.12$$

**Fig. 2.20 – First-Order Model
Development based on Existing Dynamics** 28

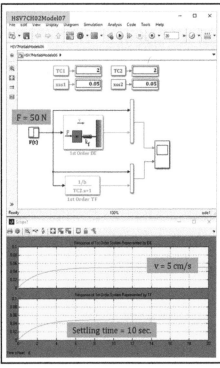

- F= 50 N.
- Settling Time = 10 s.
- Time Constant = 2 s.
- v_{SS} = 0.05 m/s (5 cm/s).

**Fig. 2.21 – Simulation
Results for the Case Study** 29

2.7- Frequency Response Analysis of First-Order Systems
2.7.1- Identification of First-Order Systems Based on Frequency Response

$$x(t) = B \sin(\omega t + \varphi) \qquad 2.13$$

$$B = \frac{A}{\sqrt{1 + \omega^2 \tau^2}} \qquad 2.14$$

$$\varphi = \tan^{-1}(-\omega \tau) \qquad 2.15$$

$$AR\,(dB) = 20\log\frac{B}{A} = 20\log\frac{1}{\sqrt{1 + \omega^2 \tau^2}} \qquad 2.16$$

30

30

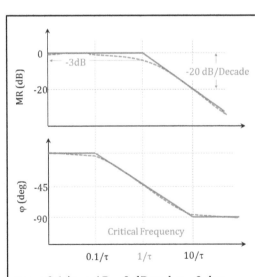

Fig. 2.22 – Bode Plot of a 1st Order System

- $\omega < 0.1/\tau \rightarrow AR \approx 0$ dB and $\varphi = 0$ deg.
- $\omega = 1/\tau$ = Critical Frequency $\rightarrow AR$ = -3 dB (called Bandwidth) and φ = -45 deg.
- $\omega >=$ Critical Frequency $\rightarrow AR$ decays 20 dB/decade of exciting frequency.
- $\omega = 10/\tau \rightarrow AR$ = -20 dB.
- $\omega => 10/\tau \rightarrow \varphi$ = -90 deg.

31

31

2.7.2- Effect of Design Parameters on 1st System Frequency Response
2.7.2.1- Effect of Time Constant

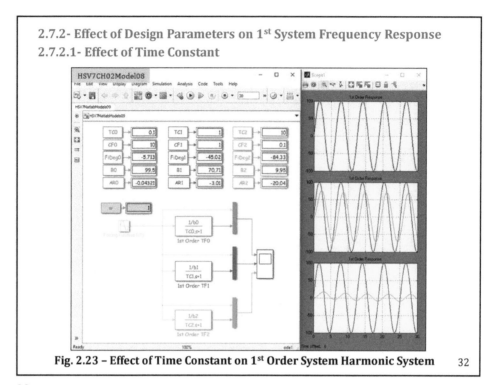

Fig. 2.23 – Effect of Time Constant on 1st Order System Harmonic System 32

32

2.7.2.2- Effect of Exciting Frequency

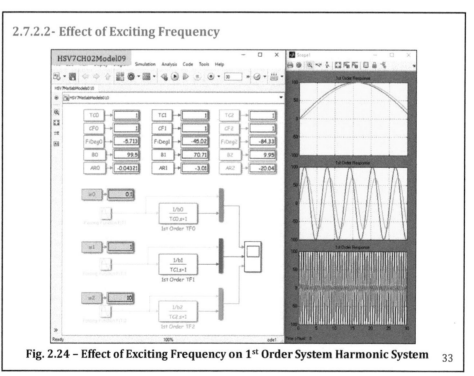

Fig. 2.24 – Effect of Exciting Frequency on 1st Order System Harmonic System 33

33

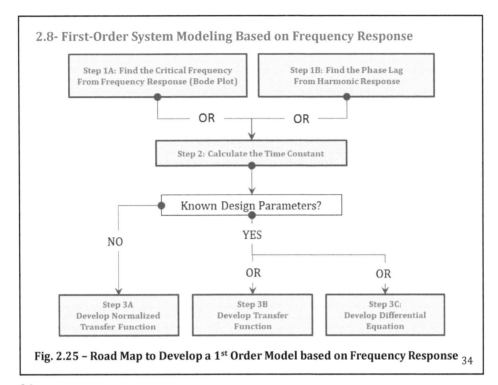

2.8- First-Order System Modeling Based on Frequency Response

Fig. 2.25 – Road Map to Develop a 1ˢᵗ Order Model based on Frequency Response 34

34

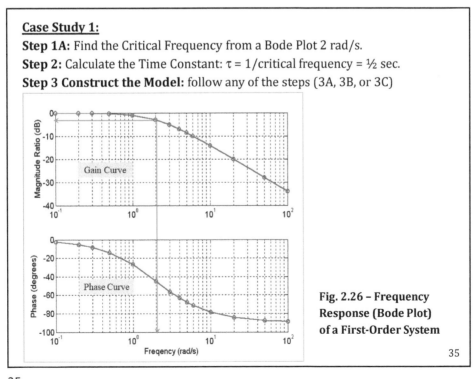

Case Study 1:

Step 1A: Find the Critical Frequency from a Bode Plot 2 rad/s.

Step 2: Calculate the Time Constant: τ = 1/critical frequency = ½ sec.

Step 3 Construct the Model: follow any of the steps (3A, 3B, or 3C)

Fig. 2.26 – Frequency Response (Bode Plot) of a First-Order System

35

35

Case Study 2:

Step 1B: Find the Phase Lag φ from Harmonic Response:

- Complete cycle time = 2 sec. $\rightarrow$ exciting frequency ω= ½ Hz. = 3.14 rad/s.
- Phase lag = -0.25 sec out of =2 sec. Since the
- Complete cycle = 360 degrees $\rightarrow$ phase lag φ = (-0.25/2) 360 = -45 deg.

Step 2: Calculate the Time Constant:

- Eq. 2.15 $\rightarrow \tau$ = - [tan (φ)]/ω.= - [tan (-45)]/3.14 = 0.516.

Step 3 Construct the Model: follow any of the steps (3A, 3B, or 3C)

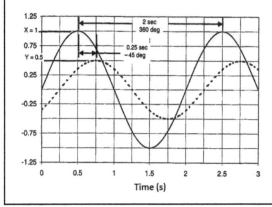

Fig. 2.27 – Harmonic Response of a First-Order System

36

36

Chapter 2 Reviews

1. For a 1ˢᵗ order represented by the equation "$k_fv + k_sx = F(t)$", time constant equal?
   A. v/x
   B. x/v
   C. k_s/k_f
   D. k_f/k_s

2. To make the 1ˢᵗ order (shown above in question 1) more responsive?
   A. Decrease the time constant
   B. Decrease the coefficient of friction, k_f
   C. Increase the spring constant, k_s
   D. All the above

3. For the 1ˢᵗ order (shown above in question 1), magnitude of the displacement **x** equals the magnitude of the input force **F** only if?
   A. $k_f = 1$
   B. $k_s = 1$
   C. $v = 1$
   D. $x = 1$

4. For a 1ˢᵗ order represented by the equation "$ma + k_fv = F(t)$", time constant equal?
   A. v/m
   B. F/v
   C. m/k_f
   D. k_f/m

5. To make the 1ˢᵗ order (shown above in question 4) less responsive?
   A. Decrease the time constant
   B. Increase the coefficient of friction, k_f
   C. Increase the mass, **m**
   D. All the above

6. For the 1ˢᵗ order (shown above in question 4), magnitude of the velocity **v** equals the magnitude of the input force **F** only if?
   A. $k_f = 1$
   B. $m = 1$
   C. $v = 1$
   D. $a = 1$

7. For a 1st order that settles after 10 seconds, time constant equals?
   A. 20 seconds
   B. 2 seconds
   C. 0.2 seconds
   D. 0.02 seconds

8. For a 1st order that has a time constant = 0.1 sec., the exciting frequency at which the system lags by 45-degree equals?
   A. 0.1 rad/s
   B. 1 rad/s
   C. 10 rad/s
   D. 100 rad/s

9. For a 1st order system that is excited by a sinusoidal signal with critical frequency, amplitude ratio of the system response equals?
   A. 70%
   B. – 3dB
   C. Bandwidth
   D. All the above

10. For a 1st order system that is excited by a sinusoidal signa, with the increase of the exciting frequency?
   A. Amplitude ratio decreased
   B. Amplitude ratio increased
   C. Bandwidth increased
   D. Bandwidth decreased

Chapter 2 Assignment

Student Name: --- Student ID: ------------------

Date: --- Score: ------------------------

Given: The hydraulic cylinder, shown below in the figure, has an equivalent moving mass **m = 2000 kg**. The cylinder has been extended suddenly. Dynamics of the cylinder has been captured. The cylinder was found performing like a first-order system and reached a constant velocity in 10 seconds.

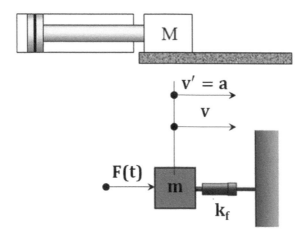

1- Develop the Normalized Transfer Function that represents the dynamic pattern of the system.
2- Develop the Transfer Function that represents the system.
3- Develop the Differential Equation that represents the system.

Chapter 3
Modeling and Simulation of Second-Order Dynamic Systems

Objectives:

In this chapter, methods and theories presented in Chapter 1 are applied for Second-Order dynamic system. The chapter presents mathematical modeling for second-order systems in t-domain and s-domain. This chapter also presents the response of second-order systems to the typical forcing functions including, step, ramp, and harmonic inputs. The chapter discusses measured characteristics of second-order step response and how to develop the transfer function of the system based on existing dynamic characteristics.

0

0

Brief Contents:

3.1- Second-Order Physical System Identification

3.2- Second-Order System Mathematical Modeling

3.3- Simulation Model Development of Second-Order Systems

3.4- Performance Simulation of Second-Order Systems

3.5- Step Response Analysis of Second-Order Systems

3.6- Second-Order System Modeling Based on Step Response

3.7- Frequency Response Analysis of Second-Order Systems

3.8- Second-Order System Modeling Based on Frequency Response

1

1

3.1- Second-Order Physical System Identification

Second-Order system contains three elements
(inductive, resistive, and capacitive).

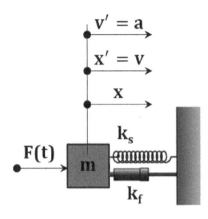

Fig. 3.1 – Mechanical Translational Second-Order Systems

2

2

3.2- Second-Order System Mathematical Modeling
3.2.1- Second-Order System Mathematical Model in t-domain

Three physical elements (inductive, resistive, and capacitive),

❖ For a Mechanical Translational System:

$$F(t) = m\frac{dv}{dt} + k_f\,v + k_s\int v\,dt \qquad\qquad 3.1$$

❖ For a Mechanical Rotational System:

$$T(t) = j\frac{d\omega}{dt} + k_f\,v + k_s\int \omega\,dt \qquad\qquad 3.2$$

❖ For an Electrical System:

$$e(t) = L\frac{di}{dt} + R\,i + \frac{1}{C}\int i\,dt \qquad\qquad 3.3$$

❖ For a Hydraulic System:

$$\Delta P(t) = I\frac{dQ}{dt} + R\,v + \frac{1}{C}\int Q\,dt \qquad\qquad 3.4$$

❖ Eq. 3.1 → $ma + k_f v + k_s x = F(t)$ \qquad 3.5

3

3

Applying Laplace Transform to D.E. 3.1→,

$$(ms^2 + k_f s + k_s)\, x(s) = F(s) \qquad\qquad 3.6$$

$$\left(s^2 + \frac{k_f}{m}s + \frac{k_s}{m}\right) x(s) = \frac{F(s)}{m} \qquad\qquad 3.7$$

$$\textbf{Undamped Natural Frequency} = \omega_n[\text{rad/s}] = \sqrt{\frac{k_s}{m}} \qquad 3.8$$

$$\textbf{Damping Ratio} = \xi = \frac{k_f}{2\sqrt{k_s m}} \qquad\qquad 3.9$$

$$TF = \frac{\textbf{Output}}{\textbf{Input}} = \frac{x(s)}{F(s)} = \frac{1/m}{s^2 + 2\xi\omega_n s + \omega_n^2} \qquad 3.10$$

4

4

To calculate the steady state value in s-domain, the Final Value Theorem applies as follows:

$$x_{ss} = \lim_{s\to 0} s.\, x(s) = \lim_{s\to 0} \frac{sF(s)/m}{s^2 + 2\xi\omega_n s + \omega_n^2} \qquad 3.11$$

$$\textbf{Step Input} \to F(s) = F/s \to$$

$$x_{ss} = \lim_{s\to 0} \frac{s\frac{F}{s}/m}{s^2 + 2\xi\omega_n s + \omega_n^2} = \frac{F}{m\omega_n^2} = \frac{F}{k_s}$$

"Hook's Law"

5

5

3.2.3- Second-Order System Normalized Transfer Function

Parameter **c** in the generic equation 1.2 (Spring constant k_s in Eq. 3.5 = 1)

$\rightarrow$ When k_s = 1, Eq. 3.8 $\rightarrow$ 1/m = ω^2 $\rightarrow$

then Eq. 3.10 can be rewritten in the following NTF

$$NTF = \frac{\textbf{Output}}{\textbf{Input}} = \frac{x(s)}{F(s)} = \frac{\omega_n^2}{s^2 + 2\xi\omega_n s + \omega_n^2} \qquad 3.12$$

$k_f = 1, k_s = 1$, step input $F(t) = 100$.

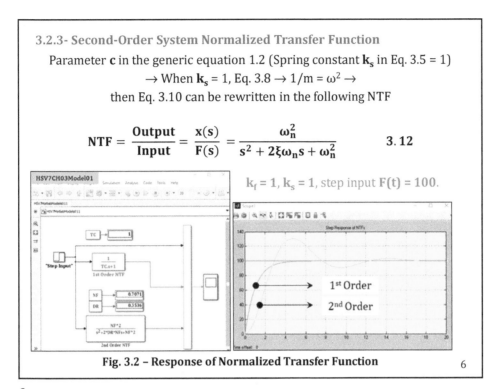

Fig. 3.2 – Response of Normalized Transfer Function

6

6

3.3- Simulation Model Development of Second-Order Systems

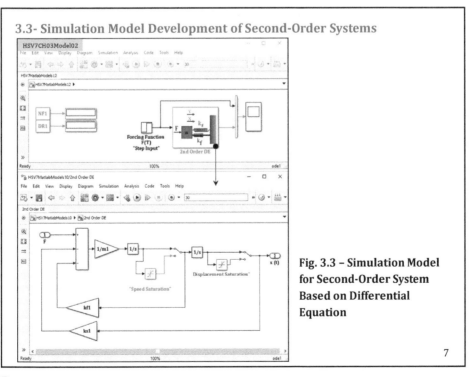

Fig. 3.3 – Simulation Model for Second-Order System Based on Differential Equation

7

7

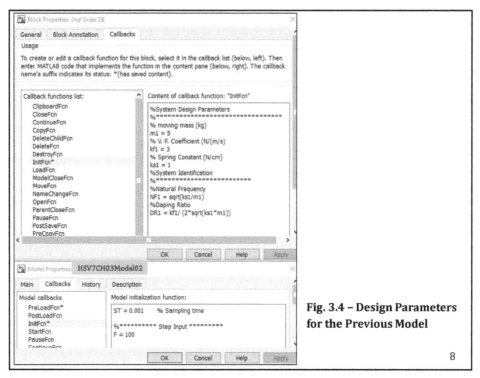

Fig. 3.4 – Design Parameters for the Previous Model

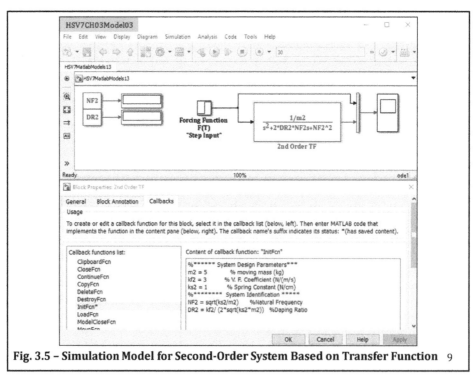

Fig. 3.5 – Simulation Model for Second-Order System Based on Transfer Function 9

3.4- Performance Simulation of Second-Order Systems

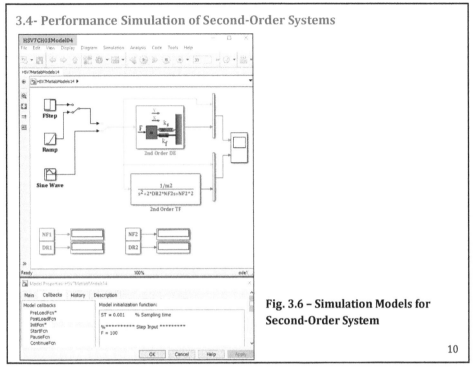

Fig. 3.6 – Simulation Models for Second-Order System

10

10

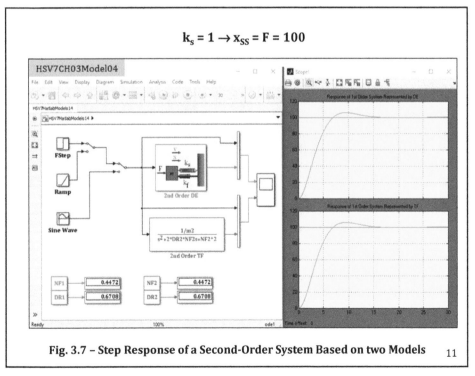

$$k_s = 1 \rightarrow x_{SS} = F = 100$$

Fig. 3.7 – Step Response of a Second-Order System Based on two Models 11

11

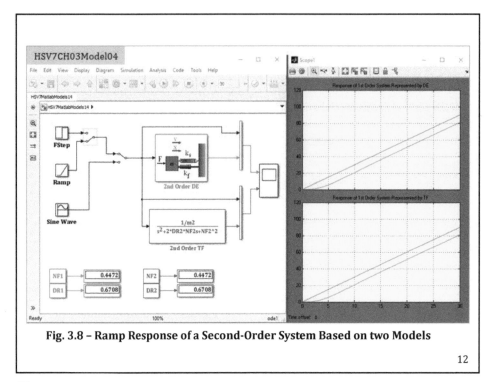

Fig. 3.8 – Ramp Response of a Second-Order System Based on two Models

12

12

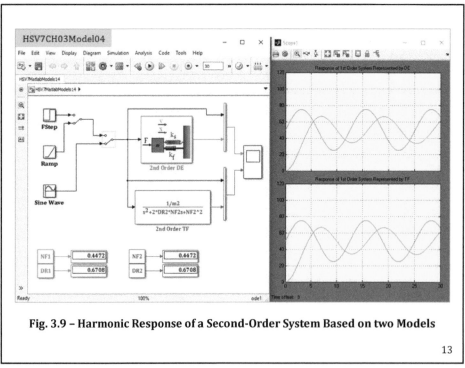

Fig. 3.9 – Harmonic Response of a Second-Order System Based on two Models

13

13

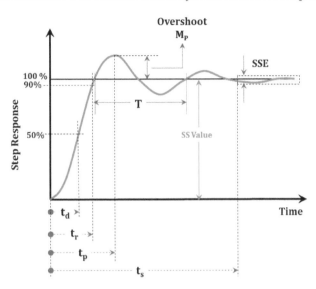

3.5- Step Response Analysis of Second-Order Systems
3.5.1- Identification of Second-Order System Based on Step Response

Fig. 3.10 – Step Response Analysis for Second-Order Systems   14

14

For underdamped 2nd order systems ($0 < \xi < 1$)	
Natural Frequeny [Hz] $= f_n = \dfrac{1}{T[s]}$	3.13
Natural Frequeny [rad/s] (defined in Eq. 3.8) $= \omega_n = 2\pi f_n$	3.14
Rise Time [s] $= t_r \approx 1.8/\omega_n$	3.15
Peak Time [s] $= t_P = \dfrac{\pi}{\omega_n\sqrt{1-\xi^2}}$	3.16
Settling Time [s] $= t_s = \dfrac{4.6}{\xi\omega_n}$ for SSE $= \pm 1\%$	3.17A
Settling Time [s] $= t_s = \dfrac{4}{\xi\omega_n}$ for SSE $= \pm 2\%$	3.17B
Settling Time [s] $= t_s = \dfrac{3}{\xi\omega_n}$ for SSE $= \pm 5\%$	3.17C 15

15

$$\text{Overshoot} = M_P = SS\ e^{\left[\frac{-\pi\xi}{\sqrt{1-\xi^2}}\right]} \qquad 3.18$$

$$\text{Percentage Overshoot} = M_P\,(\%) = 100(M_P/SS = 100\ e^{\left[\frac{-\pi\xi}{\sqrt{1-\xi^2}}\right]} \qquad 3.19$$

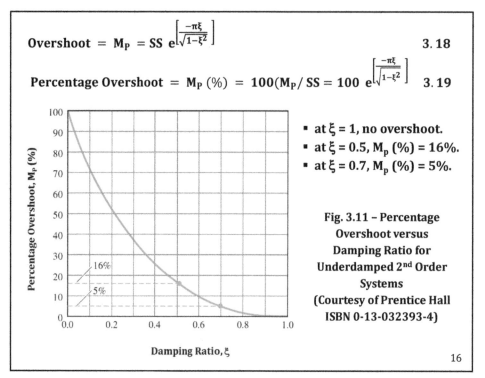

- at ξ = 1, no overshoot.
- at ξ = 0.5, M_p (%) = 16%.
- at ξ = 0.7, M_p (%) = 5%.

Fig. 3.11 – Percentage Overshoot versus Damping Ratio for Underdamped 2ⁿᵈ Order Systems (Courtesy of Prentice Hall ISBN 0-13-032393-4)

16

16

To verify the set of previous equations, assume (m = 1, kf = 0.5, and ks = 1)

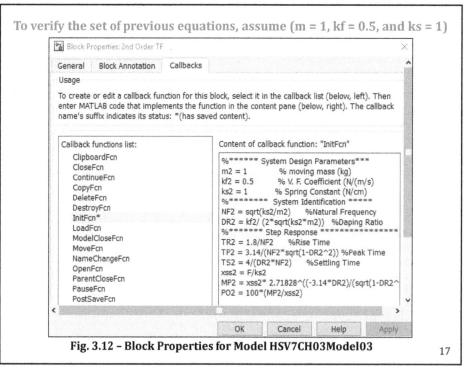

Fig. 3.12 – Block Properties for Model HSV7CH03Model03

17

17

- Eq. 3.8 $\rightarrow = \omega_n[\text{rad/s}] = \sqrt{k_s/m} = 1$
- Simulation Results (Fig. 3.13) $\rightarrow$ T = 6.28 s.
- Eq. 3.13 $\rightarrow f_n$ [Hz] = 1/T = 1/6.28.
- Eq. 3,14 $\rightarrow \omega_n$ [rad/s] = 2 πf_n = 1 that verifies Eq. 3.8.
- Eq. 3.9 $\rightarrow = \xi = [k_f/2\sqrt{k_s m}] = 0.5/2 = 0.25$
- Since spring constant = 1, then steady state value = F = 100 (see Fig. 3.13).
- Eq. 3.15 $\rightarrow$ Rise Time = 1.8/1 = 1.8 s (see Fig. 3.13).
- Eq. 3.16 $\rightarrow$ Peak Time = 3.24 s (see Fig. 3.13).
- Eq. 3.17B $\rightarrow$ Settling Time = 16 s (see Fig. 3.13).
- Eq. 3.18 $\rightarrow$ Overshoot = 44.45 (see Fig. 3.13).
- Eq. 3.19 $\rightarrow$ % Overshoot = 44.45% (see Fig. 3.13).
- Fig. 3.11$\rightarrow$ at ξ = 0.25. % Overshoot = 44.45 that verifies Eq.3.19.

18

18

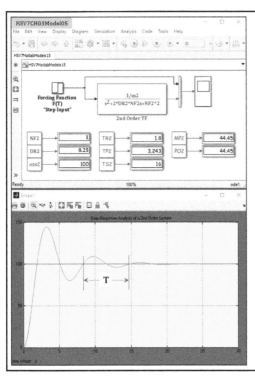

**Fig. 3.13 – Verification of a
Step Response of a Second
Order System**

19

19

3.5.2- Effect of Design Parameters on 2nd-Order System Step Response
3.5.2.1- Effect of Moving Mass

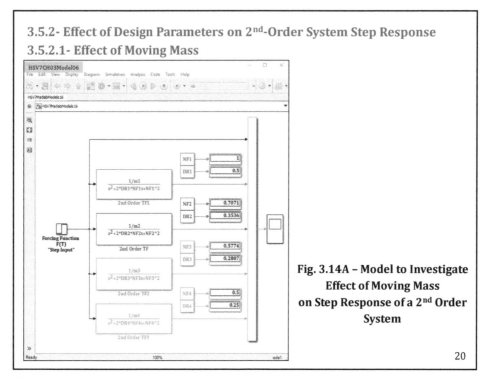

Fig. 3.14A – Model to Investigate Effect of Moving Mass on Step Response of a 2nd Order System

20

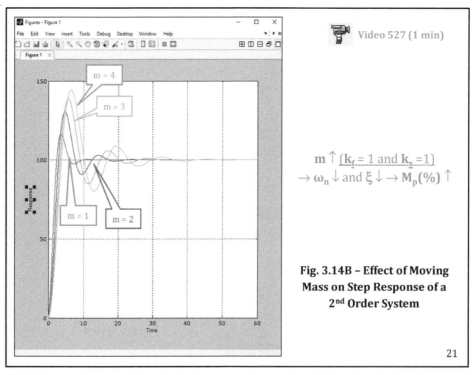

Video 527 (1 min)

$m \uparrow \underline{(k_f = 1 \text{ and } k_s = 1)}$
$\rightarrow \omega_n \downarrow \text{ and } \xi \downarrow \rightarrow M_p(\%) \uparrow$

Fig. 3.14B – Effect of Moving Mass on Step Response of a 2nd Order System

21

3.5.2.2- Effect of Friction Coefficient

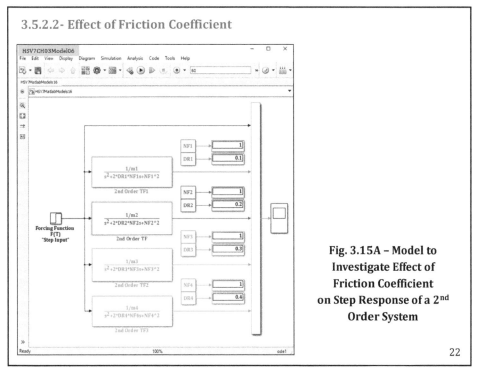

Fig. 3.15A – Model to Investigate Effect of Friction Coefficient on Step Response of a 2ⁿᵈ Order System

22

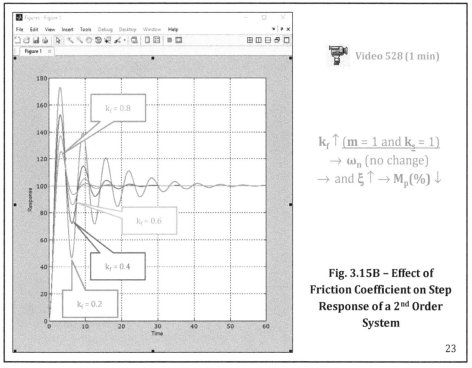

Video 528 (1 min)

$k_f \uparrow \underline{(m = 1 \text{ and } k_s = 1)}$
$\rightarrow \omega_n$ (no change)
$\rightarrow$ and $\xi \uparrow \rightarrow M_p(\%) \downarrow$

Fig. 3.15B – Effect of Friction Coefficient on Step Response of a 2ⁿᵈ Order System

23

3.5.2.3- Effect of Spring Constant

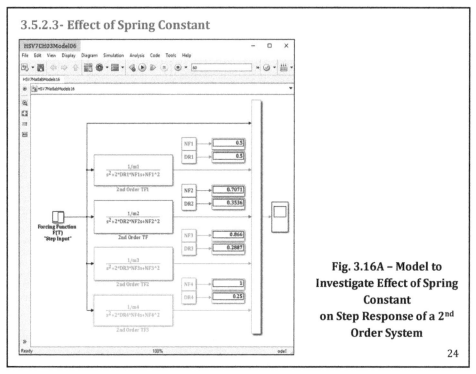

Fig. 3.16A – Model to Investigate Effect of Spring Constant on Step Response of a 2nd Order System

24

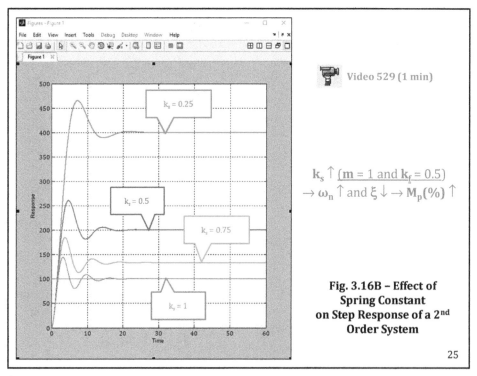

Video 529 (1 min)

$k_s \uparrow \underline{(m = 1 \text{ and } k_f = 0.5)} \rightarrow \omega_n \uparrow \text{ and } \xi \downarrow \rightarrow M_p(\%) \uparrow$

Fig. 3.16B – Effect of Spring Constant on Step Response of a 2nd Order System

25

3.5.2.4- Effect of Natural Frequency

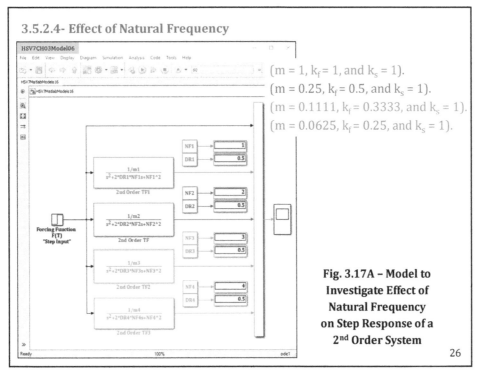

$(m = 1, k_f = 1, \text{ and } k_s = 1)$.

$(m = 0.25, k_f = 0.5, \text{ and } k_s = 1)$.

$(m = 0.1111, k_f = 0.3333, \text{ and } k_s = 1)$.

$(m = 0.0625, k_f = 0.25, \text{ and } k_s = 1)$.

Fig. 3.17A – Model to Investigate Effect of Natural Frequency on Step Response of a 2nd Order System

26

26

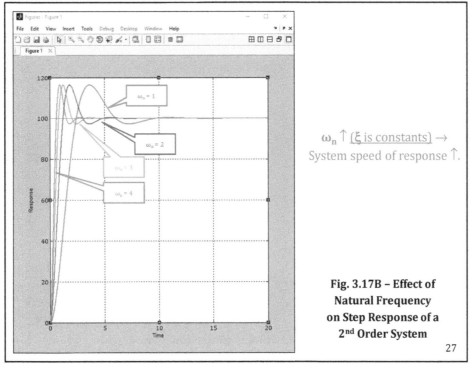

$\omega_n \uparrow (\underline{\xi \text{ is constants}}) \rightarrow$
System speed of response $\uparrow$.

Fig. 3.17B – Effect of Natural Frequency on Step Response of a 2nd Order System

27

27

3.5.2.5- Effect of Damping Ratio

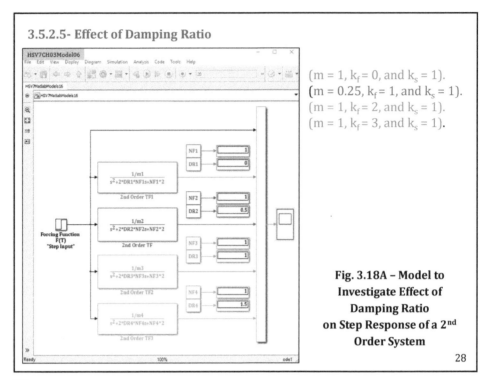

$(m = 1, k_f = 0, \text{ and } k_s = 1)$.
$(m = 0.25, k_f = 1, \text{ and } k_s = 1)$.
$(m = 1, k_f = 2, \text{ and } k_s = 1)$.
$(m = 1, k_f = 3, \text{ and } k_s = 1)$.

Fig. 3.18A – Model to Investigate Effect of Damping Ratio on Step Response of a 2nd Order System

28

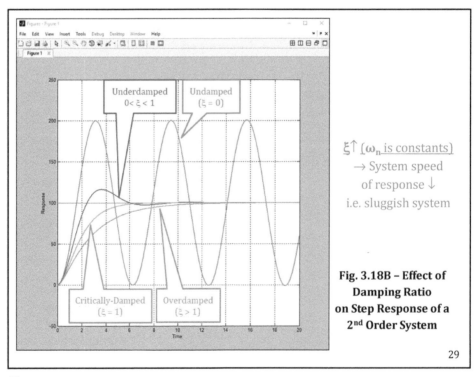

$\xi \uparrow$ (ω_n is constants)
$\rightarrow$ System speed
of response $\downarrow$
i.e. sluggish system

Fig. 3.18B – Effect of Damping Ratio on Step Response of a 2nd Order System

29

3.5.2.6- Difference between 1st Order and Overdamped 2nd Order Step Response

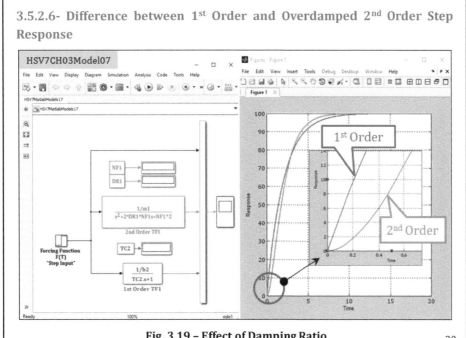

Fig. 3.19 – Effect of Damping Ratio

30

30

3.5.2.7- Effect of Damping Ratio as Presented on s-Plane

Characteristic Equation of a 2nd order system where **r** is called the roots.

$$r^2 + 2\xi\omega_n r + \omega_n^2 = 0 \qquad 3.20$$

The general solution of Eq. 3.20 results in

$$r_{1,2} = -\xi\omega_n \pm \omega_n\sqrt{(\xi^2 - 1)} \; = -\xi\omega_n \pm j\omega_n\sqrt{(1 - \xi^2)} \qquad 3.21$$

Where **j** is an imaginary number $= \sqrt{-1}$

31

31

Case 1: Unstable System ($\xi < 0$):

$$\xi < 0 \rightarrow r_{1,2} = -\xi\omega_n \pm j\omega_n = \text{+ve Real Part} \pm \text{Imaginary Part}$$

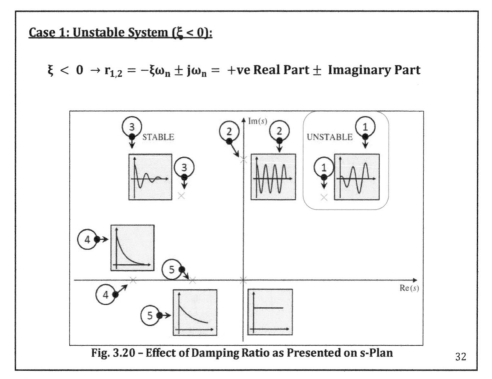

Fig. 3.20 – Effect of Damping Ratio as Presented on s-Plan

32

32

Case 2: Undamped System ($\xi = 0$) is Marginally Stable:

$$\xi = 0 \rightarrow r_{1,2} = \pm j\omega_n = \pm \text{Imaginary Part}$$

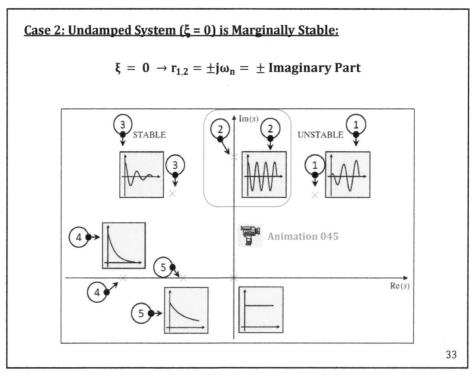

33

33

Case 3: Underdamped System ($0 < \xi < 1$) is Stable:

$$0 < \xi < 1 \rightarrow r_{1,2} = -\xi\omega_n \pm j\omega_n\sqrt{(1-\xi^2)}$$
$$= -\text{ve Real Part} \pm \text{ Imagenary Part}$$

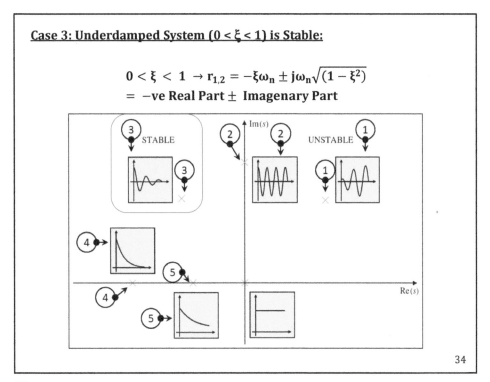

34

34

Case 4: Critically Damped System ($\xi = 1$) is Stable:

$$\xi = 1 \rightarrow r_1 = r_2 = -\xi\omega_n = -\text{ve Real Part (Coincided Roots)}$$

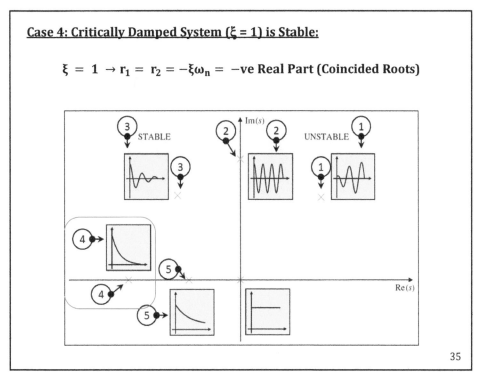

35

35

Case 5: Overdamped System ($\xi > 1$) is Stable:

$$\xi > 1 \rightarrow r_{1,2} = -\xi\omega_n \pm \omega_n\sqrt{(\xi^2 - 1)} = -ve \text{ Real Part (seprated roots)}$$

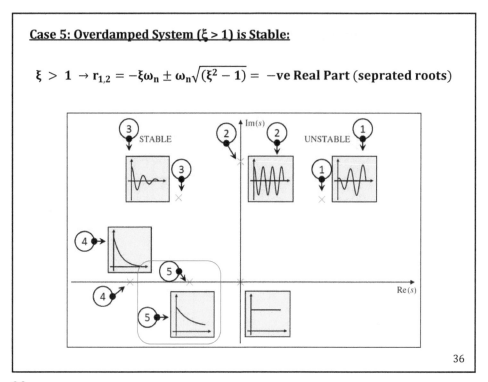

36

36

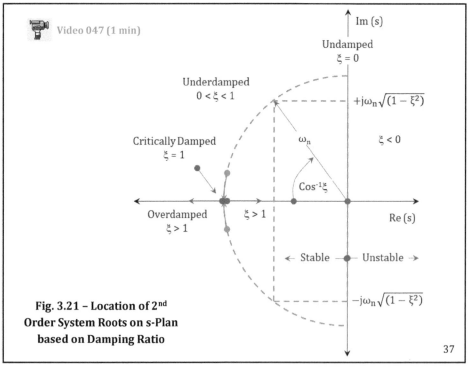

Fig. 3.21 – Location of 2nd Order System Roots on s-Plan based on Damping Ratio

37

3.6- Second-Order System Modeling Based on Step Response

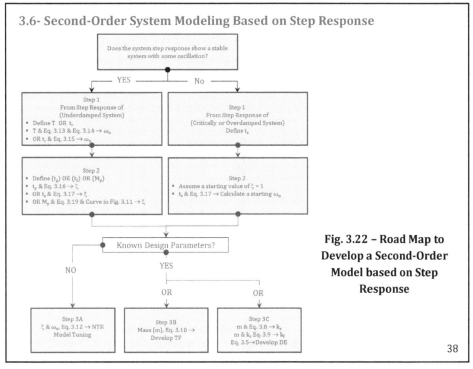

Fig. 3.22 – Road Map to Develop a Second-Order Model based on Step Response

38

38

Case Study 1 (Critically or Overdamped 2nd Order System ξ >=1):

Given: A step response of proportional directional valve
- No oscillation, settling time = 18 ms, Spool stroke = 10 mm,
- Spool mass = 100 grams, maximum EM force **F** = 122.5 N.

Step 1: Settling Time t_s: step response → Settling time (given) t_s = 18 ms.

Step 2:
- **Damping Ratio ξ:** Assume a starting value of Damping Ratio ξ = 1.
- **Natural Frequency ω_n:**
- Assuming SSE = ± 1%, Eq. 3.17A → starting ω_n (rad/s) = 4.6 / 0.018 = 255.

39

39

75

Step 3A: Develop the NTF
(in cases where physical design parameters are not known)

$\xi = 1$ & ω_n is tuned to 350 rad/s to meet the required response.

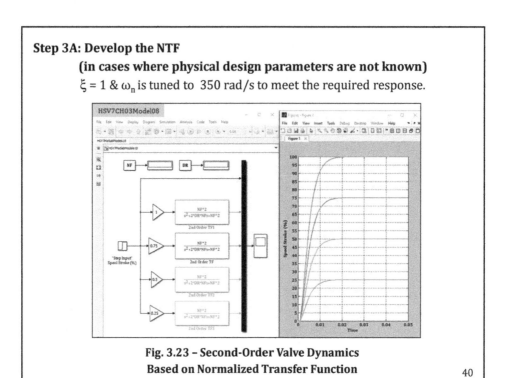

**Fig. 3.23 – Second-Order Valve Dynamics
Based on Normalized Transfer Function**

40

40

Step 3B: Develop the TF
(in cases where physical design parameters are known)

- Spool mass **m** = 0.1 kg is added to the model properties.

- The model calculates % of spool stroke (maximum stroke = 10 mm).

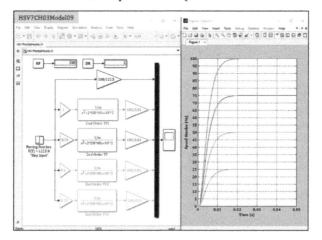

Fig. 3.24 – Second-Order Valve Dynamics Based on Transfer Function

41

41

OR Step 3C: Developing DE
(in cases where physical design parameters are known)

- **Spring Constant k_s:**
- **S**pool mass m, Eq. 3.8 $\rightarrow k_s = m\,(\omega_n)^2 = 0.1\,(350)^2 = 12250$ N/m.

- **Friction Coefficient k_f:**
- Eq. 3.9 $\rightarrow k_f = 2\xi\,(mk_s)^{0.5} = 2 \times 1\,(0.1 \times 12250)^{0.5} = 70$ kg/s.

- **Develop DE: m, $k_{f. \text{ and }} k_{s.,}$**
- Eq. 3.5 $\rightarrow$ DE as follows.

$$m[kg] \times a\left[\frac{m}{s^2}\right] + k_f\left[\frac{kg}{s}\right] \times v\left[\frac{m}{s}\right] + k_s\left[\frac{N}{m}\right] \times x\,[m] = F(t)\,[N] \qquad 3.22$$

$$0.1 \times a\left[\frac{m}{s^2}\right] + 70 \times v\left[\frac{m}{s}\right] + 12250 \times x\,[m] = F(t)\,[N] \qquad 3.23$$

42

42

The model calculates % of spool stroke (maximum stroke = 10 mm).

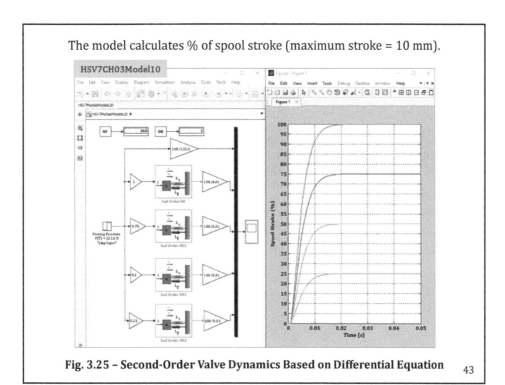

Fig. 3.25 – Second-Order Valve Dynamics Based on Differential Equation 43

43

Case Study 2 (Underdamped 2nd Order System $0 < \xi < 1$):

Given: A step response of a hydraulic actuator:

- Rise Time t_r = 0.287 s.
- Peak Time t_p = 0.51 s.
- Settling Time t_s = 3.7.
- Overshoot M_p = 53.
- Equivalent Moving Mass m = 10 kg.
- Targeted Actuator Stroke = 100 mm.

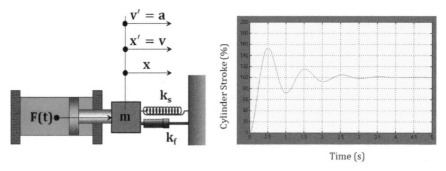

Fig. 3.26 – Experimental Step Response of a Hydraulic Actuator

44

44

Step 1- Natural Frequency ω_n:

- **Complete Cycle Time T**: T (from peak-to-peak) = 1 s.
- Eq. 3.13 → f_n [Hz] = 1/T = 1/1= 1.
- Eq. 3,14 → ω_n [rad/s] = 2 πf_n = 6.28

- **OR Rise Time t_r:** step response → $t_r \approx$ 0.287
- Eq. 3.15 → ω_n [rad/s] = = 1.8/ $t_r \approx$ 6.28

Step 2- Damping Ratio ξ:

- **Peak Time t_p:** step response → t_p = 0.51. Eq. 3.16 → ξ = 0.2

- **OR Settling Time t_s:** step response → t_s = 3.7 s. Eq. 3.17 → $\xi \approx$ 0.2

- **OR Overshoot Mp:** M_p = 53. Eq. 3.19 → M_p % = 53 & Curve (Fig. 3.11) → $\xi \approx$ 0.2

45

Step 3A: Develop the NTF
(in cases where physical design parameters are not known):

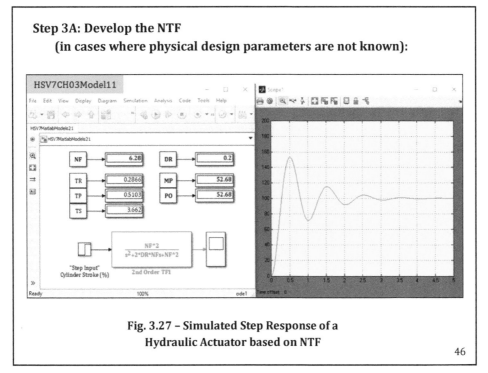

Fig. 3.27 – Simulated Step Response of a
Hydraulic Actuator based on NTF

46

46

Step 3B: Develop the TF
(in cases where physical design parameters are known):

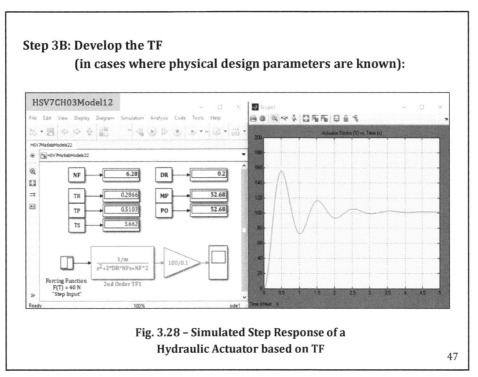

Fig. 3.28 – Simulated Step Response of a
Hydraulic Actuator based on TF

47

47

OR Step 3C: Developing DE
  (in cases where physical design parameters are known):

- **Spring Constant k_s:** spool mass m, Eq. 3.8 $\rightarrow k_s = m\,(\omega_n)^2 = 10\,(6.28)^2 = 394.4$ N/m.

- **Friction Coefficient k_f:**
- Eq. 3.9 $\rightarrow k_f = 2\xi\,(mk_s)^{0.5} = 2 \times 0.2\,(10 \times 394.4)^{0.5} = 25.12$ kg/s.

- **Develop DE:** knowing **m, k_f and k_s**
- Eq. 3.5 $\rightarrow$ DE as follows.

$$m[kg] \times a\left[\frac{m}{s^2}\right] + k_f\left[\frac{kg}{s}\right] \times v\left[\frac{m}{s}\right] + k_s\left[\frac{N}{m}\right] \times x\,[m] = F(t)\,[N] \qquad 3.22$$

$$10 \times a\left[\frac{m}{s^2}\right] + 25.12 \times v\left[\frac{m}{s}\right] + 394.4 \times x\,[m] = F(t)\,[N] \qquad 3.23$$

48

48

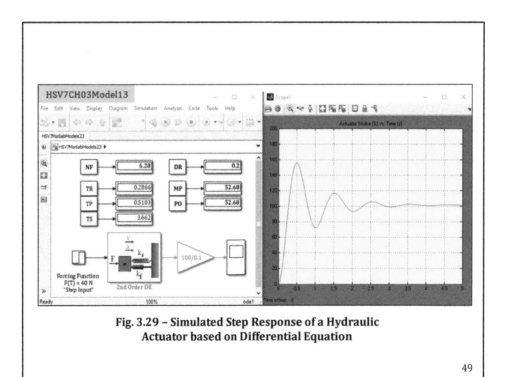

**Fig. 3.29 – Simulated Step Response of a Hydraulic
Actuator based on Differential Equation**

49

49

3.7- Frequency Response Analysis of Second-Order Systems
3.7.1- Identification of 2nd Order Systems Based on Frequency Response

$$x(t) = B\sin(\omega t + \varphi) \qquad\qquad 3.24$$

$$B = A / \sqrt{\left[\frac{2\xi\omega}{\omega_n}\right]^2 + \left[1 - \frac{\omega^2}{\omega_n^2}\right]^2} \qquad\qquad 3.25$$

$$\varphi = -\tan^{-1}\left[\left(\frac{2\xi\omega}{\omega_n}\right) / \left(1 - \frac{\omega^2}{\omega_n^2}\right)\right] \qquad\qquad 3.26$$

$$AR\,(dB) = 20\log\frac{B}{A} = 20\log\left[1 / \sqrt{\left[\frac{2\xi\omega}{\omega_n}\right]^2 + \left[1 - \frac{\omega^2}{\omega_n^2}\right]^2}\right] \qquad 3.27$$

50

50

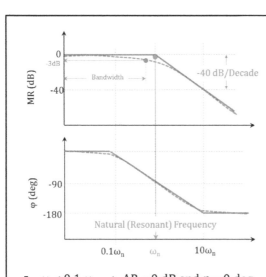

Fig. 3.30 – Bode Plot of a 2nd Order System

- $\omega < 0.1\,\omega_n \rightarrow AR \approx 0$ dB and $\varphi = 0$ deg.
- $\omega =$ Bandwidth $\rightarrow AR = -3$ dB.
- $\omega = \omega_n \rightarrow \varphi = -90$ deg.
- $\omega > \omega_n \rightarrow AR$ decays 40 dB/decade of exciting frequency.
- $\omega > 10\,\omega_n \rightarrow \varphi = -180$ deg.

51

51

3.7.2- Effect of Design Parameters on Second-Order System Frequency Response

3.7.2.1- Effect of Damping Ratio

3.7.2.2- Effect of Exciting Frequency

Same effect as on a 1st Order system.

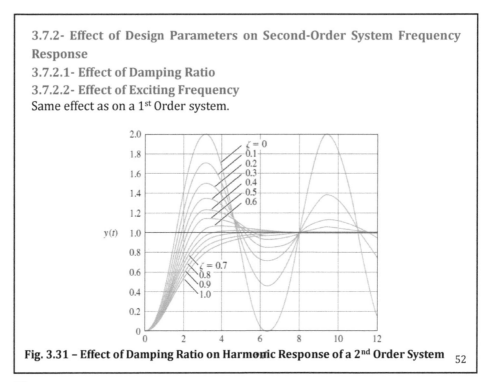

Fig. 3.31 – Effect of Damping Ratio on Harmonic Response of a 2nd Order System 52

52

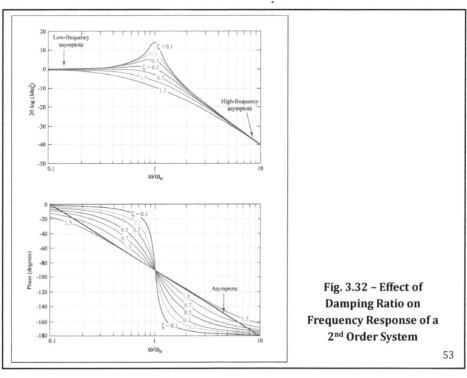

Fig. 3.32 – Effect of Damping Ratio on Frequency Response of a 2nd Order System

53

53

3.7.3- Stability Analysis Based on Bode Plot

❖ **Definitions:**

- **Gain Crossover Frequency (ω_{cg}):** When **AR** starts to descend below 0 dB.
- **Phase Crossover Frequency (ω_{cp}):** When φ first reaches (-180).
- **Gain Margin (GM):** Is the value of **AR** (at ω_{cp}).
- **Phase Margin (PM):** Is the value of φ (at ω_{cg}).

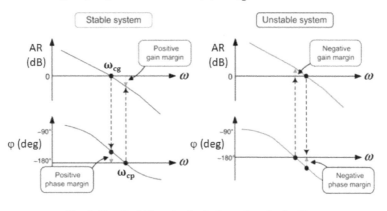

Fig. 3.33 – Stability Analysis Based on Bode Plot

54

54

3.7.3- Stability Analysis Based on Bode Plot

❖ **Stable System:**

- GM = +ve (value by which the gain curve would be lifted to reach unity).
- PM = +ve (value by which the phase curve would be lowered to reach -180).
- For a system to at least underdamped, both GM and PM must be positive.
- Large GM and large PM means highly damped and sluggish system.
- Small GM and Small PM means underdamped and responsive system.
- Typically, dynamic systems designed with GM = 10 dB and PM = 45°.

❖ **Marginally Stable System:** GM = 0 and PM = 0.

❖ **Unstable System:** GM = -ve and PM = -ve.

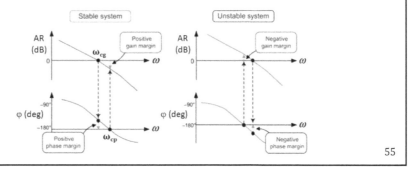

55

55

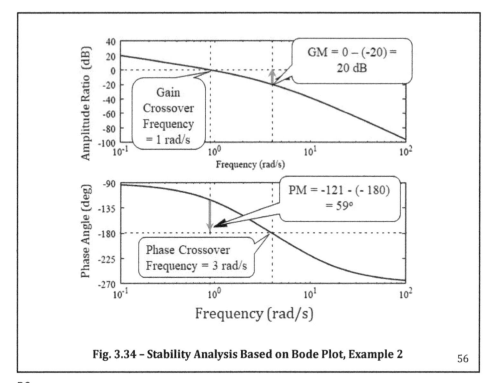

Fig. 3.34 – Stability Analysis Based on Bode Plot, Example 2

56

56

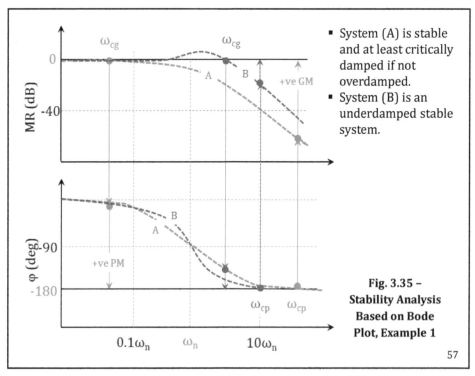

- System (A) is stable and at least critically damped if not overdamped.
- System (B) is an underdamped stable system.

Fig. 3.35 – Stability Analysis Based on Bode Plot, Example 1

57

57

3.8- Second-Order System Modeling Based on Frequency Response

Step 1- Natural Frequency ω_n:
Find the natural frequency and phase angle = -90 deg.

Step 2- Damping Ratio ξ:

❖ **For Underdamped Systems:**
If the frequency response of a 2nd order system shows that the underdamped response, calculate the % overshoot then use the Curve in Fig 13 → ξ.

❖ **For Critically Damped and Overdamped Systems:**
Assume a starting value of Damping Ratio ξ = 1 that can be tuned later. results.

❖ **Developing a Model:**
Once ξ and ω_n are calculated, based on the available physical parameters, follow the appropriate step to continue building the model as it has been discussed previously.

58

58

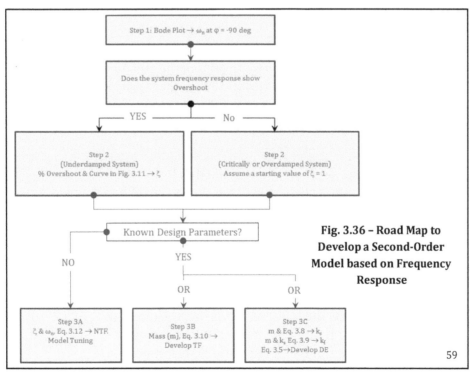

Fig. 3.36 – Road Map to Develop a Second-Order Model based on Frequency Response

59

59

Chapter 3 Reviews

1. For a 2nd order represented by the equation "$ma + k_f v + k_s x = F(t)$", if the magnitude of the mas m = coefficient of friction k_f = spring constant k_s = 2, natural frequency equal?
    A. 0.1 rad/s
    B. 1 rad/s
    C. 10 rad/s
    D. 100 rad/s

2. For the same system presented in question 1, damping ratio equal?
    A. 0.5
    B. 1
    C. 2
    D. 4

3. For the 2nd order (shown above in question 1), magnitude of the displacement **x** equals the magnitude of the input force **F** only if?
    A. kf = 1
    B. ks = 1
    C. v = 1
    D. x = 1

4. For a 2nd order system, rise time is defined as?
    A. The time the system takes to rise to 60% of the steady state value
    B. The time the system takes to rise to 70% of the steady state value
    C. The time the system takes to rise to 80% of the steady state value
    D. The time the system takes to rise to 90% of the steady state value

5. For a 2nd order system, settling time is defined as?
    A. The time the system takes to settle within a predefined steady state error, typically 1-5 %, around the steady state value.
    B. The time the system takes to settle 100% of the steady state value
    C. The time the system takes to rise to 90% of the steady state value
    D. The time the system takes to rise to 80% of the steady state value

6. Which statement reflects the effect of increasing the mass ($m\uparrow$) of a 2nd order system?
    A. Natural Frequency $\downarrow$, Damping Ratio $\downarrow$, % Overshoot $\downarrow$
    B. Natural Frequency $\uparrow$, Damping Ratio $\downarrow$, % Overshoot $\uparrow$
    C. Natural Frequency $\downarrow$, Damping Ratio $\downarrow$, % Overshoot $\uparrow$
    D. Natural Frequency $\downarrow$, Damping Ratio $\uparrow$, % Overshoot $\uparrow$

7. Which statement reflects the effect of increasing the friction coefficient ($k_f\uparrow$) of a 2nd order system?
   A. Natural Frequency ↓, Damping Ratio ↓, % Overshoot ↓
   B. Natural Frequency not affected, Damping Ratio ↑ % Overshoot ↓
   C. Natural Frequency ↑, Damping Ratio ↑, % Overshoot ↑
   D. Natural Frequency ↓, Damping Ratio ↑, % Overshoot not affected

8. Which statement reflects the effect of increasing the spring constant ($k_s\uparrow$) of a 2nd order system?
   A. Natural Frequency ↓, Damping Ratio ↓, % Overshoot ↓
   B. Natural Frequency ↑, Damping Ratio ↓, % Overshoot ↑
   C. Natural Frequency ↓, Damping Ratio ↓, % Overshoot ↑
   D. Natural Frequency ↓, Damping Ratio ↑, % Overshoot ↑

9. A 2nd order system that has a damping ration greater than zero and less than 1 is considered as?
   A. Undamped system
   B. Underdamped system
   C. Critically damped system
   D. Overdamped system

10. Natural frequency of a 2nd order system is the frequency at which the system lags behind the exciting frequency by?
    A. 45 degrees
    B. 90 degrees
    C. 120 degrees
    D. 180 degrees

Chapter 3 Assignment

Student Name: --- Student ID: ------------------

Date: -- Score: ------------------------

Assignment: Us the model # (HSV7CH03Model05) to find the response parameter of a 2nd order that has the following design parameters:

m = 1 kg
k$_f$ = 2 [N/(m/s)]
k$_s$ = 20 (N/cm)

Chapter 4
Modeling Approaches
for Hydraulic Components and Systems

Objectives:

This chapter explores the different approaches when modeling a hydraulic component versus modeling a hydraulic system at large. The chapter presents the basic idea and the structure of lumped modeling, an adopted modeling approach for application engineers.

0

0

Brief Contents:

4.1- Modeling Approaches for Component Developers

4.2- Modeling Approaches for Application Engineers

4.3- Lumped Modeling Approach for Application Engineers

1

1

4.1- Modeling Approaches for Component Developers

4.1.1- Basic Purposes of Modeling for Component Developers

- Model: kinematics, static and dynamic performance of the component.
- Investigate: effect of the operating conditions on component performance.
- Optimize: a component design to makes it more competitive.

4.1.2- Features of Modeling for Component Developers

- Deep understanding: component construction and principle of operation.
- Access to classified information: dimensions, tolerances, material used, mass, friction coefficients, spring constants, etc.
- Complex models: lots of computational effort.
- Multiple models: kinematics, dynamics, stress, thermal, CFD, cavitation, etc.
- Customized software: manufacturer's know-how. + expensive
- Models are dedicated for a specific component/size/brand: are not applicable for other components or used universally.

2

2

4.2- Modeling Approaches for Application Engineers

4.2.1- Basic Purposes of Modeling for Application Engineers

- Study interaction between subsystems: engine, transmission, hydraulic functions, etc.
- Investigate machine performance: kinematic & dynamic.
- Investigate machine operation: duty cycles, energy efficiency, reliability, maintenance protocols, effect of hydraulic fluid, etc.

4.2.2- Features of Modeling for Application Engineers
- Machine Builders:
- They are the end users of hydraulic components.
- They select component from a wide range of manufacturers.
- They are not concerned about the design and construction of components.
- They are concerned about component performance, cost, efficiency, lifetime, operational conditions, sensitivity to contaminations, noise, etc.

3

3

- Example: using a pressure-compensated pump in an excavator, does it mean anything whether the pump is a swashplate or a vane pump?

- Therefore, machine builders adopt simplified modeling approaches for the components that form the system for the following reasons:

➢ Avoid containing large number of equations.
➢ Reduce the model complexity and the required computational effort.
➢ Avoid asking for Classified information.
➢ Focus more on overall machine performance.
➢ Components should be modeled for ease of connecting to other upstream and downstream components.

4

4

4.3- Lumped Modeling Approach for Application Engineers

4.3.1- Introduction to Lumped Modeling Approach

Conventional Approach for Simple Systems:

- Set of equations to be solved instantaneously.
- Model dedicated for a specific system.
- If the system layout is changed, the whole set of equations will be changed.

Software Platform that have Models for Generic Hydraulic Components:

- Matlab-Simhydraulics, Automation Studio, Easy 5, etc.
- Modelers have access to recharacterize the components
- Modelers have no access to change the match under the mask of the model.

Lumped modeling approach :

- System model composed of lumped component models hooked together.
- Models are based on characteristics published by the manufacturers.
- Component models can be re-characterized and used repeatedly.
- Structured with predefined input/output to easy connection.

5

5

Lumped models of hydraulic components can be classified based on the energy as:

- Energy Source: e.g. pump.
- Energy Sink: e.g. reservoir.
- Energy Storage: e.g. accumulator.
- Energy Transmission: e.g. conductor.
- Energy Consumers: e.g. actuators and valves

Lumped models can be classified based on the direction of flow as follows:

- *Irreversible* Flow Components: e.g. pressure relief valves.
- *Reversible* Flow Components: e.g. double acting cylinder. This model can be a combination of two irreversible flow components.

6

6

4.3.2- Lumped Model Structure for Hydraulic Components

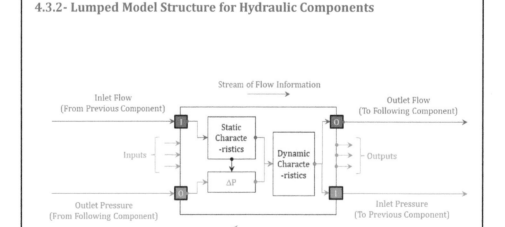

Fig. 4.1- Structure of a Lumped Model for a Hydraulic Component

7

7

4.3.3- Lumped Model Structure for Open Circuits

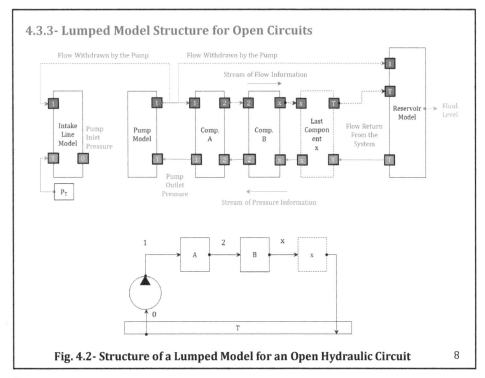

Fig. 4.2- Structure of a Lumped Model for an Open Hydraulic Circuit   8

8

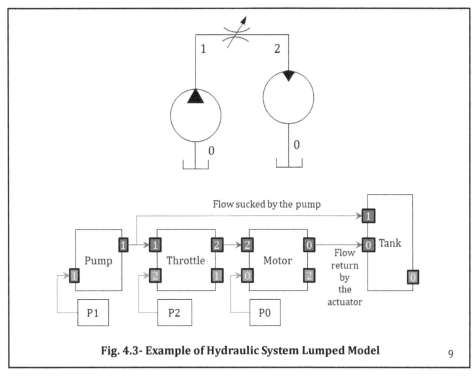

Fig. 4.3- Example of Hydraulic System Lumped Model   9

9

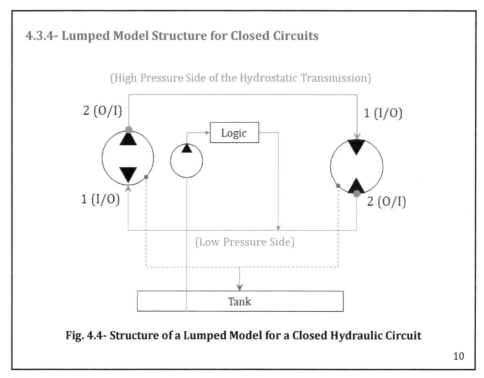

4.3.4- Lumped Model Structure for Closed Circuits

Fig. 4.4- Structure of a Lumped Model for a Closed Hydraulic Circuit

10

10

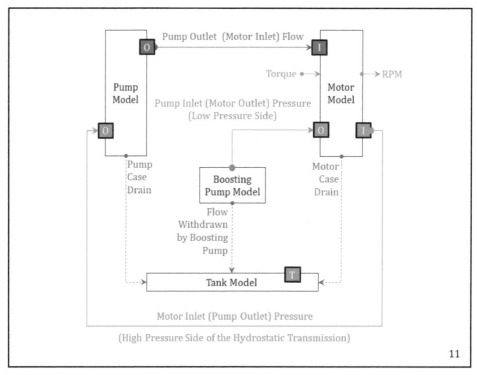

11

11

4.3.5- Simulating the Static Characteristics of a Hydraulic Component in a Lumped Model

Complexity:

- Avoid unnecessary information to reduce the complexity of the model.
- The amount of information needed should be enough to get the predefined outputs.
- Depending on the nature of the hydraulic circuit operating conditions, various levels of the model can be built.

Format:

- Static characteristics of hydraulic components are either reported by the manufacturer or developed experimentally.
- It can be found in tabulated or graphical forms.

12

12

Linear vs. Nonlinear Characteristics:

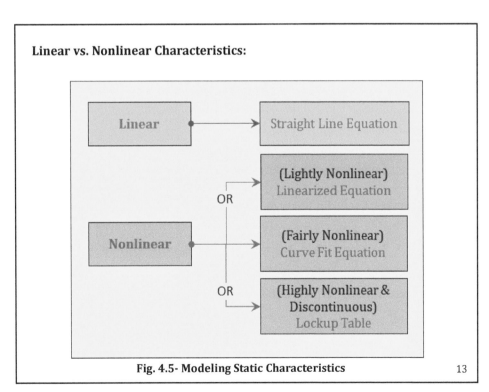

Fig. 4.5- Modeling Static Characteristics   13

13

Boundaries:
- Manufacturers test components within ranges of operating conditions.
- Reported characteristics may not cover operating conditions.
- Extrapolation may be needed beyond the reported characteristics.
- Extrapolation must always kept be within the allowable boundaries.

Example 1:
- Pump flow is measured versus a range of 100 bar.
- Maximum operating pressure reported by the manufacturer is 200 bar.
- Q-P characteristics of the pump are loaded as a lookup table.
- Any extrapolation process should not exceed 200 bar.

Example 2:
- An equation $Q = f\,(\Delta P)^{0.5}$ is used to model a directional valve.
- This equation isn't valid for all ranges of differential pressure.
- This equation isn't valid beyond the valve power limit.

14

14

Multidimensional Static Characteristics:
- In some cases, a static characteristic depends on multiple variables.
- It can be approximated as a unidimensional characteristic.
- Consider the variable that has the dominant effect and ignore the other.

Multidimensional Lookup Table:
- Useful for nonlinear multidimensional characteristics.
- Example: pump flow = f (p and n).

Taylor's Series: Useful for linear or lightly nonlinear multidimensional chs.

$$z\,(x\,\&\,y) \;=\; z_0 \;+\; \left[\frac{\delta z}{\delta x}\right]_y (x - x_0) + \left[\frac{\delta z}{\delta y}\right]_x (y - y_0) + \textbf{Nonlinear Terms} \quad \textbf{4.1}$$

So, linearized Taylor's Series is as follows:

$$z\,(x\,\&\,y) \;=\; z_0 \;+\; \left[\frac{\delta z}{\delta x}\right]_y (x - x_0) + \left[\frac{\delta z}{\delta y}\right]_x (y - y_0) \qquad \textbf{4.2}$$

An example of this is fluid viscosity is function of pressure and temperature. 15

15

4.3.6- Simulating the Dynamic Characteristics of a Hydraulic Component in a Lumped Model

Static Characteristics + Normalized Transfer Function

4.3.7- Example of Lumped Modeling

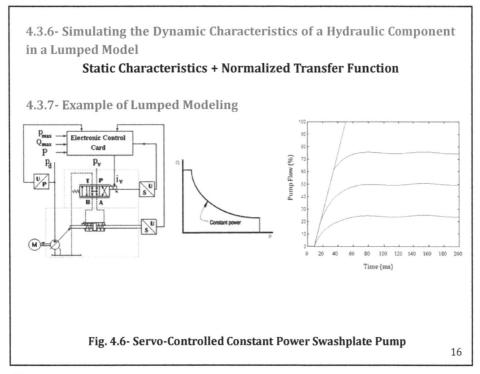

Fig. 4.6- Servo-Controlled Constant Power Swashplate Pump

16

16

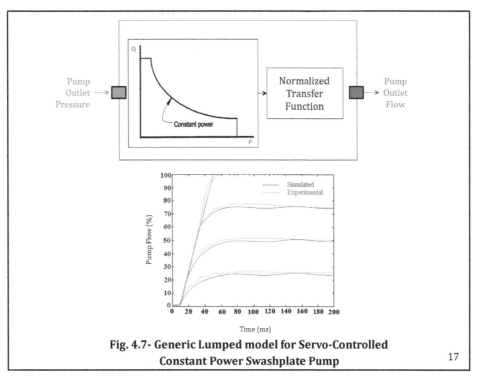

Fig. 4.7- Generic Lumped model for Servo-Controlled Constant Power Swashplate Pump

17

17

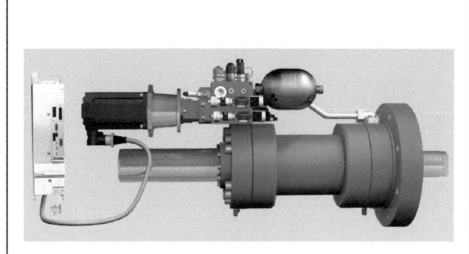

Fig. 4.8- Self-Contained Electro-Hydraulic Actuator
(Courtesy of Bosch Rexroth

18

18

Chapter 4 Reviews

1. Featured of Modeling for Component Developers?
    A. Requires access to classified information such as constructional dimensions, tolerances, material used, mass, friction coefficients, spring constants, etc.
    B. Complex models that require lots of computational effort.
    C. Multiple models may be needed, one for kinematic and dynamic analysis, one for stress analysis, one for thermal analysis, one for cavitation analysis, etc.
    D. All of the above

2. Lumped modeling as applied for a hydraulic pump is suitable for?
    A. Use of the pump in a hydraulic system of an excavator
    B. Investigate the effect of the temperature on the pump internal leakage
    C. Investigate the effect of pressure on the lifetime of pump bering surfaces
    D. All of the above

3. Detailed modeling as applied for a hydraulic pump is suitable for?
    E. Optimizing the design of a pump
    F. Investigate the effect of the temperature on the pump internal leakage
    G. Investigate the effect of pressure on the lifetime of pump bering surfaces
    H. All of the above

4. Correct technique to model highly nonlinear static characteristics is
    A. Straight line equation
    B. Lookup table
    C. Linearized equation
    D. Curve fit

5. In the lumped modeling technique presented in this textbook?
    A. Flow information flows forward and pressure information flowing backward.
    B. Flow information flows forward and pressure information flowing forward.
    C. Flow information flows backward and pressure information flowing backward.
    D. Flow information flows backward and pressure information flowing forward.

Chapter 4 Assignment

Student Name: -- Student ID: ------------------

Date: -- Score: -----------------------

Assignment: Explain, with examples, the boundaries in modeling static characteristics of a hydraulic component lumped model.

Chapter 5
Modeling of Fluid Properties

Objectives:

This chapter presents different techniques to model hydraulic fluid properties based on available information. Properties considered in this chapter are bulk modulus, density, specific gravity and viscosity. In modeling such properties, effects of working temperature and pressure are considered. Case studies are presented, and Matlab-Simulink models were built and validated based on given information.

0

0

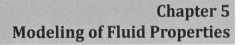

Brief Contents:

5.1- Introduction to Fluid Properties Modeling

5.2- Modeling of Hydraulic Fluid Bulk Modulus

5.3- Modeling of Hydraulic Fluid Density and Specific Gravity

5.4- Modeling of Hydraulic Fluid Viscosity

5.5- Lumped Model for Hydraulic Fluid Properties

1

1

5.1- Introduction to Fluid Properties Modeling

- Fluid properties are never maintained constant during machine operation.
- Unlike modeling a hydraulic component, the fluid properties model is not connected to upstream and downstream components.
- Modeled fluid properties are assigned to a global variables.

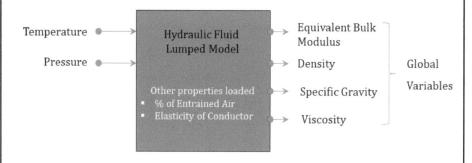

5.1- Generic Lumped Model of Fluid Properties

2

5.2- Modeling of Hydraulic Fluid Bulk Modulus
5.2.1- Definition and Mathematical Expression of Bulk Modulus

Definition:
- A property that indicates how stiff a fluid is.
- Bulk Modulus is the reciprocal of compressibility.
- The higher the bulk modulus, the stiffer and less compressible the fluid is.

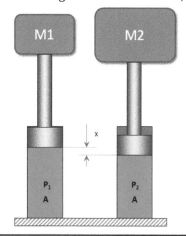

Fig. 5.2- Hydraulic Fluid Bulk Modulus vs. Compressibility

3

Mathematical Expression:

- Bulk Modulus, modulus of elasticity, & spring constant of a spring.

- All expressed as: (effort variable/reactional change).

- Modulus of Elasticity = tensile stress/ longitudinal change in the specimen.

- Spring Constant = compressive or tensile force/longitudinal change.

- Bulk Modulus = Pressure/ volumetric change of the oil volume.

$$\beta = - \frac{\Delta P}{(\frac{\Delta V}{V_o})} \qquad\qquad 5.1$$

4

4

Fluid Type	Bulk Modulus at 20 °C and 10,000 psi
Water-Glycol	500,000 psi
Water-in Oil Emulsion	333,000 psi
Phosphate Ester	440,000 psi
ISO 32 Mineral Oil	260,000 psi

Table 5.1- Bulk Modulus of Typical Hydraulic Fluids

COMPRESSIBILITY-VOLUME CHANGE (at 100 bar = 1450 psi)	
Fluid Type	% ΔV Reduction
Mineral Oil	0.7%
Vegetable-based Oil	0.5%
Water and Emulsified Water-Oil	0.4%
Water-Glycol and Synthetic Fluids (Polymers)	0.35%

Table 5.2- Compressibility of Typical Hydraulic Fluids

5

5.2.2- Case Study 1 for Modeling of Bulk Modulus
5.2.2.1- Effect of Temperature and Pressure on Bulk Modulus

- **BM** increase of working temperature
- (decreases linearly ≈10% for every 40% °F)

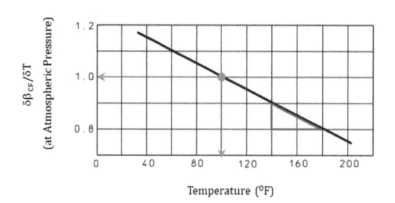

**Fig. 5.3- Pressure and Temperature Correction Factors for MIL-H-83282
Hydraulic Fluid Bulk Modulus (Courtesy of the Lee Company)**

6

6

- **BM** slightly increases with increase of working pressure.
- (increases linearly ≈10% for every 2000 psi)

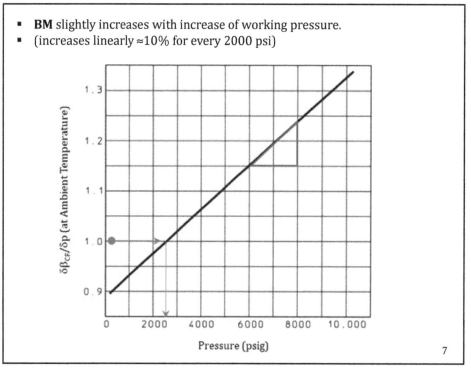

7

7

104

- This is a multidimensional linear static characteristic
- $\rightarrow$ Linearized Taylor's Series

$$\beta_{TP} = \beta_o + \left[\frac{\delta\beta}{\delta T}\right]_P (T - T_o) + \left[\frac{\delta\beta}{\delta p}\right]_T (P - P_o) \qquad 5.2$$

Fig. 5.3 doesn't show the rate of change of bulk modulus with temperature and pressure, it rather shows the *Correction Factors*. Then:

$$\beta_{TP} = \beta_o + \beta_o \left[\frac{\delta\beta_{CF}}{\delta T}\right]_P (T - T_o) + \beta_o \left[\frac{\delta\beta_{CF}}{\delta P}\right]_T (P - P_o) \Rightarrow$$

$$\beta_{TP} = \beta_o \left[1 + \left[\frac{\delta\beta_{CF}}{\delta T}\right]_P (T - T_o) + \left[\frac{\delta\beta_{CF}}{\delta P}\right]_T (P - P_o)\right] \qquad 5.3$$

- β_{TP} = Bulk modulus corrected based on current **T** and current **P**.
- β_o = 300k psi = Reference BM at reference T_o [100 ^{O}F] and P_o. [= 2500 psig].
- $(\delta\beta/\delta T)_P$ = Rate of change of β with respect to **T** at constant **P**.
- $(\delta\beta/\delta P)_T$ = Rate of change of β with respect to **P** at constant **T**.
- Given Characteristics $\rightarrow (\delta\beta_{CF}/\delta T)_P$ = (-0.1/40) = 0.0025 [1/^{O}F].
- Given Characteristics $\rightarrow (\delta\beta_{CF}/\delta P)_T$ = (0.1/2000) = 0.00005 [1/psi].

8

8

5.2.2.2- Effect of Entrained Air on Bulk Modulus

- **BM** is drastically decreased with entrained air.
- **BM** slightly affected by transmission line elasticity.

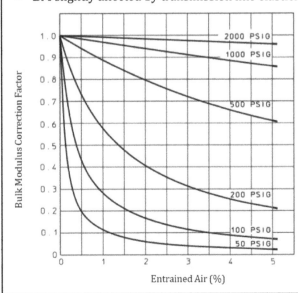

Fig. 5.4- Effect of Entrained Air on Hydrocarbon Hydraulic MIL-H-83282 Fluid Bulk Modulus (Courtesy of the Lee Company)

9

9

Equivalent BM based on air content and elasticity of transmission line walls.

$$\frac{1}{\beta_E} = \frac{1}{\beta_C} + \frac{V_F}{V_T}\frac{1}{\beta_{TB}} + \frac{V_A}{V_T}\frac{1}{\beta_A} = \frac{1}{\beta_C} + \frac{(V_T - V_A)}{V_T}\frac{1}{\beta_{TB}} + \frac{V_A}{V_T}\frac{1}{\beta_A} \qquad 5.4$$

Where:

- β_E = Effective bulk modulus
- β_C = *Modulus of Elasticity* of the transmission line (conductor).
- β_A = Air bulk modulus.
- V_F / V_T = Ratio of fluid volume to the total volume.
- V_A / V_T = Ratio of air volume to the total volume.

By defining V_{Air} as the % of air in the total volume

$$\frac{1}{\beta_E} = \frac{1}{\beta_C} + \frac{(100 - V_{Air})/100}{\beta_{TB}} + \frac{V_{Air}/100}{\beta_A} \qquad 5.5$$

10

10

Example:

- Amount of entrained air = 1%
- Oil BM= 300 x 10^3 psi (corrected based of temperature and pressure)
- Conductor bulk modulus = 200 x 10^6 psi
- Air bulk modulus = 2800 psi

Then,

$$\frac{1}{\beta_E} = \frac{1}{\beta_C} + \frac{(100 - V_{Air})/100}{\beta_{TP}} + \frac{V_{Air}/100}{\beta_A}$$

$$= \frac{1}{200 \times 10^6} + \frac{0.99}{3 \times 10^3} + \frac{0.1}{2800} = \frac{1}{1.455 \times 10^3}$$

This example shows that, increasing air content by 1% drops the oil bulk modulus from 300 x 10^3 to 1.455 x 10^3 that is equivalent to 51.5% of the fluid's bulk modulus with no air content.

11

11

5.2.2.3- Structure of Lumped Model for Case Study 1

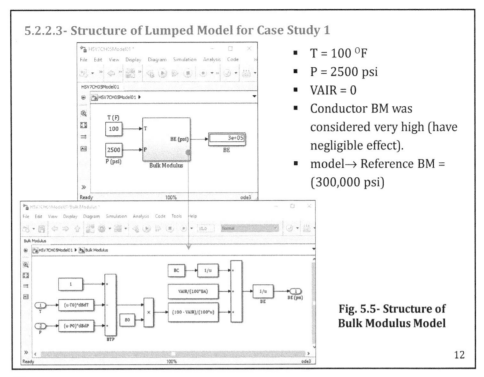

- T = 100 ^{O}F
- P = 2500 psi
- VAIR = 0
- Conductor BM was considered very high (have negligible effect).
- model→ Reference BM = (300,000 psi)

Fig. 5.5- Structure of Bulk Modulus Model

12

12

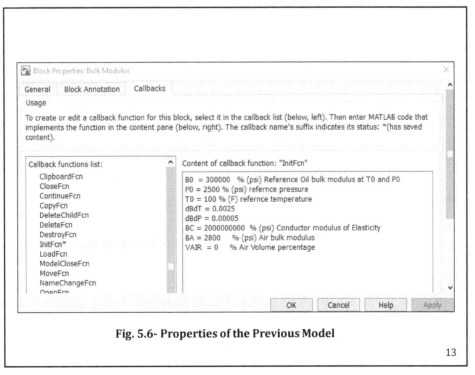

Fig. 5.6- Properties of the Previous Model

13

13

1% entrained air →
Equivalent BM decreased from 300,000 psi to 145,500 psi

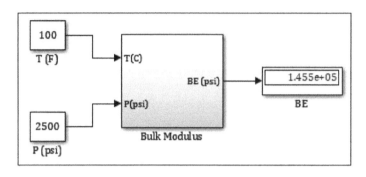

Fig. 5.7- Validating Case Study 1

14

14

5.3- Modeling of Hydraulic Fluid Density and Specific Gravity
5.3.1- Definitions and Mathematical Expression of Density

Hydraulic Fluid *Density* is defined as the amount of mass in a unit volume of the fluid. Equations 5.6 and 5.7 show the mathematical quantification of fluid density in Metric and English system of units; respectively.

$$\rho_f \left[\frac{kg_m}{m^3}\right] = \frac{Mass\ (kg_m)}{Volume\ (m^3)} \qquad 5.6$$

Where density of water at ambient conditions $\rho_w = 1000\ kg_m/m^3$

$$\rho_f \left[\frac{lb_m}{ft^3}\right] = \frac{Mass\ (lb_m)}{Volume\ (ft^3)} \qquad 5.7$$

Where density of water at ambient conditions $\rho_w = 62.4\ lb_m/ft^3$

15

15

5.3.2- Definitions and Mathematical Expression of Specific Weight

Hydraulic Fluid *Specific Weight* is defined as the weight of a unit volume of the fluid.

$$\gamma_f \left[\frac{kg_f}{m^3} \right] = \frac{Weight\ (kg_f)}{Volume\ (m^3)} = \frac{Mass\ (kg_m) \times g\ (m/s^2)}{Volume\ (m^3)} = \rho_f \times g \qquad 5.8$$

specific weight of water $\gamma_w = \rho_w$ x g = 1000 [kg_m / m^3] x 9.81 [m/s^2]
$$= 1000 \times 9.81\ [kg_m .m/\ s^2]\ [1/m^3]$$
$$= 1000 \times 9.81\ [N/m^3] = 1000\ [kg_f/m^3]$$

$$\gamma_f \left[\frac{lb_f}{ft^3} \right] = \frac{Weight\ (lb_f)}{Volume\ (ft^3)} = \frac{Mass\ (lb_m) \times g\ (ft/s^2)}{Volume\ (ft^3)} = \rho_f \times g \qquad 5.9$$

Specific weight of water $\gamma_w = \rho_w$ x g = 62.4 [lb_m / ft^3] x 32.2 [ft/s2]
$$= 62.4 \times 32.2\ [lb_m. ft/s^2]\ [1/ft^3]$$
$$= 62.4 \times 32.2\ [Poundal/ft^3] = 62.4\ [lb_f/ft^3]$$

5.3.3- Definitions and Mathematical Expression of Specific Gravity

- *Specific Gravity* is dimensionless.
- SG < 1 means it is lighter than water and vice versa.

$$SG_f = \frac{\gamma_f}{\gamma_w} = \frac{\rho_f}{\rho_w} \qquad 5.10$$

16

16

5.3.4- Case Studies for Modeling of Density and Specific Gravity
5.3.4.1- Case Study 2 based on Effect of T and P on Density

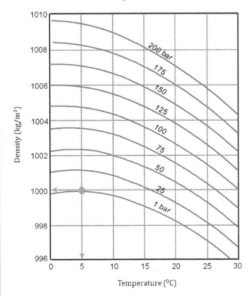

- $T \uparrow \rightarrow \rho \downarrow \& P \uparrow \rightarrow \rho \uparrow$
- To model such effects, information should be known for a specific fluid.

- **Modeling Method1:**
- Multidimensional lookup table.

Fig. 5.8- Effects of Pressure and Temperature on Fluid Density (www.engineeringtoolbox.com)

17

17

Modeling Method 2: Linearized Taylor's series.

- Reference density ρ_o = 1000 kg/m3
- ρ_o = 1000 kg/m^3 = Reference density
- At reference values T_o [0 ^{O}C = 5 ^{O}F] and P_o [14.5 psig = 1 barg].
- $(\delta\rho/\delta P)_T$ seems approximately linear $\approx$ (1/25) = 0.04 [(kg/m^3)/bar].

- Corrected density based on pressure.

$$\rho_P = \rho_o + \left[\frac{\delta\rho}{\delta P}\right]_T (P - P_o) \qquad\qquad 5.11$$

- $(\delta\rho/\delta T)_p$ is nonlinear $\rightarrow$ correction factor lookup table

T [C]	0	5	10	15	20	25	30
DT	1	1	1	0.999	0.998	0.997	0.995

Table 5.3- Temperature-Density Correction Factor Lockup Table

18

18

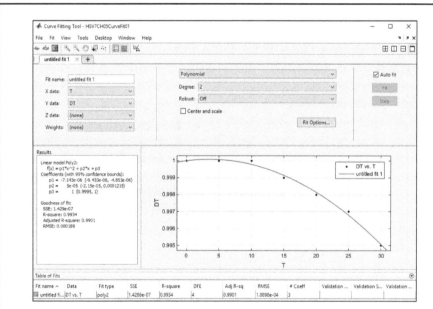

Fig. 5.9- Temperature-Density Correction Factor Curve Fit Equation

19

19

5.3.4.2- Structure of Lumped Model for Case Study 2

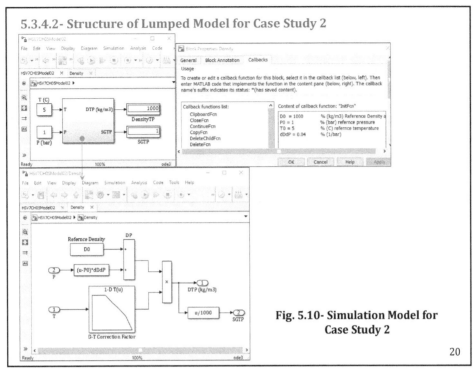

Fig. 5.10- Simulation Model for Case Study 2

20

20

5.3.4.3- Case Study 3 based on Density as Function of other Properties

linearized Taylor's series →

$$\rho_{TP} = \rho_o + \left[\frac{\delta\rho}{\delta P}\right]_T (P - P_o) + \left[\frac{\delta\rho}{\delta T}\right]_P (T - T_o) \qquad 5.12$$

Eq. 5.1 $\rightarrow \beta = -V_o \left[\frac{\delta P}{\delta V}\right]_T$ **& since** $V \uparrow \rightarrow \rho \downarrow$ **then** $\rightarrow \beta = \rho_o \left[\frac{\delta P}{\delta \rho}\right]_T \rightarrow \left[\frac{\delta P}{\delta \rho}\right]_T = \frac{\beta}{\rho_o}$

Hence, $\qquad \left[\frac{\delta\rho}{\delta P}\right]_T = \frac{\rho_o}{\beta} \qquad 5.13$

$\alpha = \frac{1}{V_o} \left[\frac{\delta V}{\delta T}\right]_P$ **& since** $V \uparrow \rightarrow \rho \downarrow$ **then** $\rightarrow \alpha = -\frac{1}{\rho_o} \left[\frac{\delta\rho}{\delta T}\right]_P$

Hence, $\qquad \left[\frac{\delta\rho}{\delta T}\right]_P = -\alpha\,\rho_o \qquad 5.14$

Substitute Equations 5.12 and 5.13 in 5.11 results in

$$\rho_{TP} = \rho_o \left[1 + \frac{(P - P_o)}{\beta_{TP}}\right] - \alpha_{TP}(T - T_o) \qquad 5.15$$

21

21

5.3.4.4- Structure of Lumped Model for Case Study 3

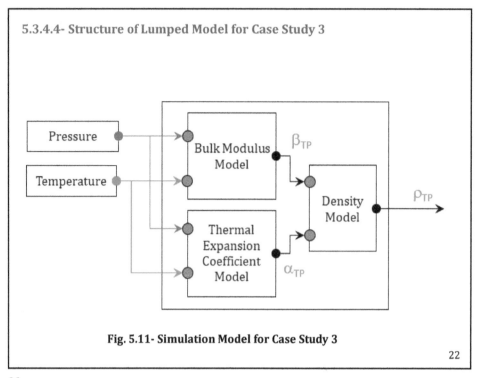

Fig. 5.11- Simulation Model for Case Study 3

22

22

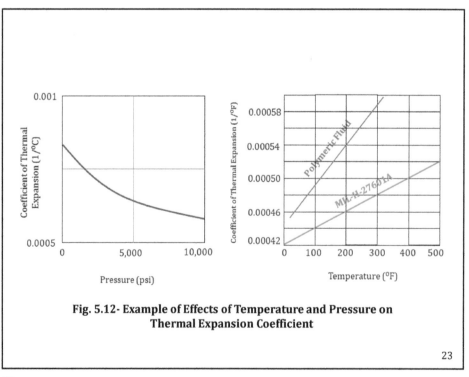

Fig. 5.12- Example of Effects of Temperature and Pressure on Thermal Expansion Coefficient

23

23

5.3.4.5- Case Study 4 based on Effect of Temperature and Pressure on Specific Gravity

$$SG_P = SG_o + \left[\frac{\delta SG}{\delta P}\right]_T (P - P_o) \quad 5.16$$

- SG_P = Density corrected based on current working pressure **p**.
- $SG_o = 1$
- At T_o [= 100 ^{o}F] and p_o [= 0 psig].
- $(\delta SG/\delta P)_T \approx (0.004/1000)$
- = 4×10^{-6} [(1/psi].

Correction based on T: The figure shows that the slope decreases consistently approximately 0.2% per every 100 ^{o}F as temperature increases.

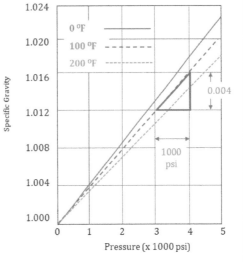

Fig. 5.13- Specific Gravity versus Pressure

24

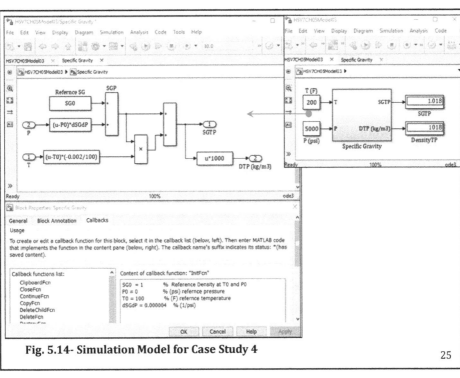

Fig. 5.14- Simulation Model for Case Study 4

25

5.4- Modeling of Hydraulic Fluid Viscosity
5.4.1- Definitions and Mathematical Expression of Viscosity

Viscosity: Resistance of the fluid to flow.

$$F \propto \frac{vA}{y} \rightarrow F = \mu \frac{vA}{y} \rightarrow \textbf{Dynamic (Absolute) Viscosity } \mu = \frac{F/A}{v/y} \qquad \textbf{5.17}$$

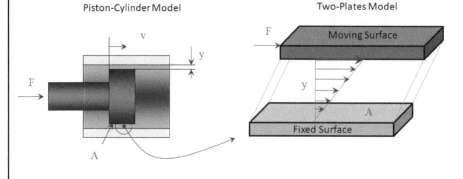

Fig. 5.15- Developing Mathematical Expression for Fluid Viscosity

26

26

Units of Fluid Viscosity are as follows:

Metric Units of Dynamic (Absolute) Viscosity $\mu = \dfrac{\text{Shear Stress}}{\text{Shear Rate}}$ is $\left[\dfrac{\text{N.s}}{\text{m}^2}\right]$.

Industrial Unit: $1cP = 10^{-3} \left[\dfrac{\text{N.s}}{\text{m}^2}\right]$.

Metric Units of Kinematic (Relative) Viscosity $= \dfrac{\mu}{\rho} =$ is $\left[\dfrac{\text{m}^2}{\text{s}}\right]$.

Industrial Unit: $1cSt = 10^{-6} \left[\dfrac{\text{m}^2}{\text{s}}\right] = \left[\dfrac{\text{mm}^2}{\text{s}}\right]$.

27

27

For balancing the units, Eq. 5.18 →

$$\text{Kinematic (Relative) Viscosity} \quad \nu\left[\frac{m^2}{s}\right] = \frac{\mu\left[\frac{N.s}{m^2}\right]}{\rho\left(\frac{kg}{m^3}\right)}$$

$$\rightarrow 10^{-6}\nu\,(cSt) = \frac{\mu\,(cP)\text{x}\,10^{-3}}{\rho\left(\frac{kg}{m^3}\right)} \rightarrow \nu\,(cSt) = \frac{1000\,\text{x}\,\mu\,(cP)}{\rho\left(\frac{kg}{m^3}\right)} = \frac{1000\,\text{x}\,\mu\,(cP)}{SG\times\rho_W\left(\frac{kg}{m^3}\right)} \qquad 5.19$$

Knowing that the water density is equal to 1000 kg/m3, equation 5.20 →

$$\mu\,(cP) = \frac{\left|\nu\,(cSt)\times SG\times\rho_W\left(\frac{kg}{m^3}\right)\right|}{1000} = \nu\,(cSt)\times SG\times\rho_W\left(\frac{g}{cc}\right) \qquad 5.20$$

- Where SG is the fluid specific gravity.
- $\mu\,(cP)$ for water =1 cP at 25 $^{\circ}$C.
- $\nu\,(cSt)$ for water = 1 cSt.

28

28

5.4.2- Case Study 5 for Modeling of Viscosity Based on Effect of Temperature and Pressure

T ↑→ Viscosity ↓ & P ↑→ Viscosity ↑

- ISO VG 100 hydraulic fluid:
- T_0 = 40 $^{\circ}$C and ν_0= 100 cSt.
- Logarithmic scale
- → $\nu - T$ isn't a linear.
- → $\nu - T$ lookup table.
- Assuming viscosity
- increases ≈1% /10 bar.

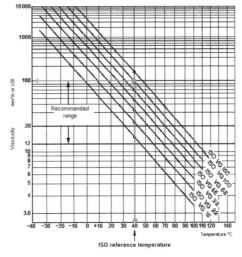

29

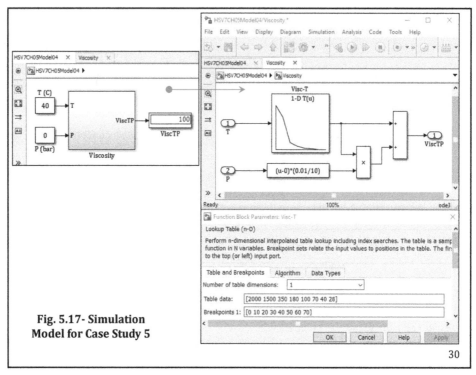

Fig. 5.17- Simulation Model for Case Study 5

30

5.5- Lumped Model for Hydraulic Fluid Properties

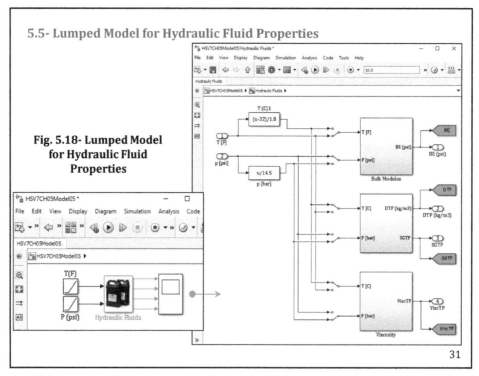

Fig. 5.18- Lumped Model for Hydraulic Fluid Properties

31

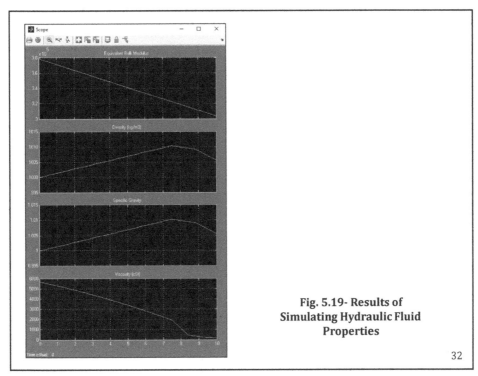

Fig. 5.19- Results of Simulating Hydraulic Fluid Properties

32

32

Chapter 5 Reviews

1. Larger Bulk Modulus means?
   A. Hydraulic fluid is more compressible
   B. Hydraulic control system will have reduced bandwidth
   C. Hydraulic actuators are sluggish
   D. All the above statements are incorrect

2. Which of the following operating conditions has the most effect on the fluid bulk modulus?
   A. Operating temperature
   B. Operating pressure
   C. Air content in the oil
   D. None of the above condition affect fluid bulk modulus

3. Specific weight of a hydraulic is defined as?
   A. Mass of a unit volume of the fluid
   B. Weight of a unit volume of the fluid
   C. Weight of a unit volume of the water
   D. Density of fluid / density of water

4. Units of specific gravity of a fluid is?
   A. kg/m^3
   B. Dimensionless
   C. lb_f/in^3
   D. lb_m/in^3

5. Which of the following has the least effect on fluid viscosity?
   A. Operating temperature
   B. Operating pressure
   C. Air content in the oil
   D. None of the above condition affect the fluid viscosity

Chapter 5 Assignment

Student Name: --- Student ID: ------------------

Date: --- Score: -----------------------

Assignment: Use model (HSV7CH05Model01) to calculate the % drop in bulk modulus due to 2% air content. All built in parameter in the model properties stay the same.

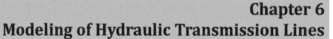

Chapter 6
Modeling of Hydraulic Transmission Lines

Objectives:

This chapter presents modeling transmission lines, fittings and orifices. Model for a transmission line considers compressible fluid so that effect of line capacitance can be investigated. Developed models were validated based on other software.

0

0

Brief Contents:

6.1- Modeling of Seamless Hydraulic Transmission Lines

6.2- Modeling of Hydraulic Fittings

6.3- Modeling of Hydraulic Orifices

6.4- Modeling Hydraulic Transmission Line Assembly

1

1

6.1- Modeling of Seamless Hydraulic Transmission Lines

1. **Lumped Resistance**: In this portion of the line, pressure drops due to fluid-fluid and fluid-wall friction, but flow will be the same as Qin.
2. **Lumped Inductance**: In this portion of the line, pressure drops due to fluid inertia, but flow will be the same as Qin.
3. **Lumped Capacitance:** In this portion of the line, outlet flow changes based on the fluid compressibility and the rate of change in the differential pressure across the line.

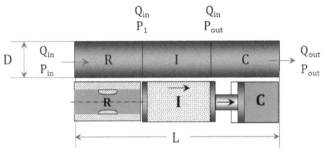

Fig. 6.1 – Modeling Pressure Drop in Hydraulic Transmission Lines   2

2

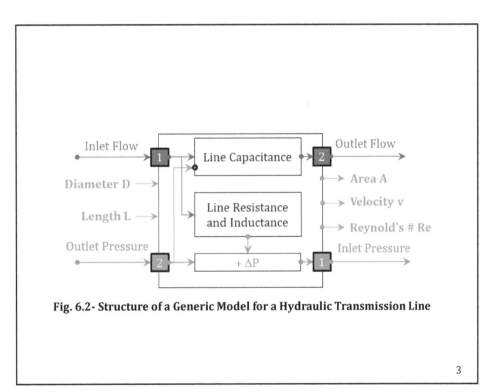

Fig. 6.2- Structure of a Generic Model for a Hydraulic Transmission Line

3

3

The challenge to build a Simulink model is to make sure units are balanced. Therefore, the following set of variables with the assigned units are used to build the model.

- ρ **(kg/m^3)** = Fluid Density (must be corrected based on working temperature and pressure).
- **v (m/s)** = Fluid Velocity.
- **D (mm)** = Line Diameter.
- **A [cm^2]** = Line Area.
- **L (m)** = Line Length.
- **R_e (-) = Reynold's Number**
- **λ (-) =** Coefficient of Fluid Friction.
- **v (cSt = mm^2/s)** = Kinematic Viscosity.
- **Q_{in} (lit/min) =** Inlet Flow.
- **Q_{out} (lit/min) =** Outlet Flow.
- **P_{in} (bar) =** Inlet Pressure.
- **P_{out}(bar) =** Outlet Pressure.

4

4

6.1.1- Modeling Pressure Losses due to Line Resistivity

Model is structured based on the sequence of the following set of equations:

$$A \left[cm^2\right] = \left[\pi \times D^2 (mm)\right] / (4 \times 100) \qquad 6.1$$

$$v \, [m/s] = \frac{Q_{in} (lit/min) \times \frac{1}{60 \times 1000}}{A \, (cm^2) \times 10^{-4}} = \frac{Q_{in} (lit/min)}{A \, (cm^2) \times 6} \qquad 6.2$$

$$R_e \, [-] = \frac{v \, [m/s] \times 1000 \times D \, [mm]}{v \, (Cst \, = \, mm^2/s)} \qquad 6.3$$

For Laminar and Transitional Flow (Reynolds's Number <= 3500):

$$\lambda = \frac{64}{R_e} \qquad 6.4$$

For Turbulent Flow (Reynolds's Number > 3500) & smooth lines:

$$\lambda = \frac{0.3164}{R_e^{0.25}} \qquad 6.5$$

5

5

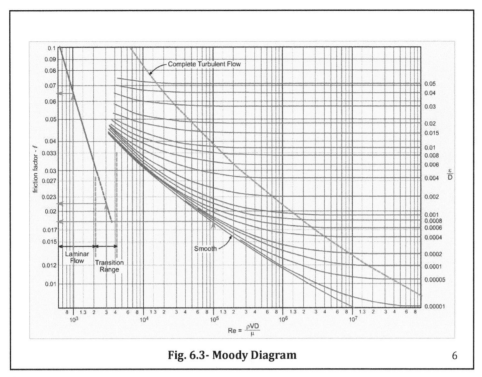

Fig. 6.3- Moody Diagram   6

6

$$\Delta P_R = \lambda \frac{L}{D} \frac{\rho v^2}{2} = \lambda \frac{L}{D} \frac{SG \times \rho_W \times v^2}{2} \qquad 6.6$$

$$\Delta P_R(\text{Pasca} = \frac{\lambda \times L(m)}{D(mm) \times 10^{-3}} \frac{SG \times 1000 \ (kg/m^3) \times v^2(m/s)^2}{2}$$

$$= \frac{10^6 \times \lambda \times SG \times L \times v^2}{2 \times D} \left[\frac{kg.m}{s^2} \frac{1}{m^2} = \frac{N}{m^2} \right]$$

$$\Rightarrow \Delta P_R(\text{bar}) = \frac{\Delta p_R(\text{Pascal})}{10^5} = \frac{5 \times \lambda \times SG \times L(m) \times [v(m/s)]^2}{D(mm)} \qquad 6.7$$

7

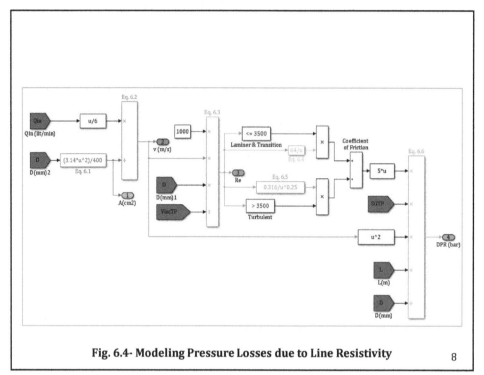

Fig. 6.4- Modeling Pressure Losses due to Line Resistivity

8

8

6.1.2- Modeling Pressure Losses due to Line Inductance

Eq. 1.6D and Eq. 1.7 in Chapter 1 $\rightarrow$ $\qquad \Delta P_I = \frac{\rho L}{A} \times \frac{dQ_{in}}{dt}$ $\qquad$ 6.8A

$$\Delta P_I \text{ (Pascal)} = \frac{\rho \left[\frac{kg}{m^3}\right] L[m]}{A[cm^2] \times 10^{-4}} \times \frac{dQ_{in}\left[\frac{lit}{min}\right] \times \frac{1}{60 \times 1000}}{dt[s]}$$

$$\Rightarrow \Delta P_I \text{ (Pascal)} = \frac{\rho L}{A \times 10^{-4}} \times \frac{dQ_{in} \times \frac{1}{60 \times 1000}}{dt} \left[\frac{kg \times m \times m^3}{m^3 \times m^2 \times s^2}\right]$$

$$\Rightarrow \Delta P_I \text{ (Pascal)} = \frac{\rho L}{A \times 6} \times \frac{dQ_{in}}{dt} \left[\frac{kg.m}{s^2} \frac{1}{m^2} = \frac{N}{m^2}\right]$$

$$\Rightarrow \Delta P_I \text{ (bar)} = \frac{\rho \left[\frac{kg}{m^3}\right] L[m]}{A[cm^2] \times 6 \times 10^5} \times \frac{dQ_{in}(lit/min)}{dt} \qquad 6.8B$$

9

9

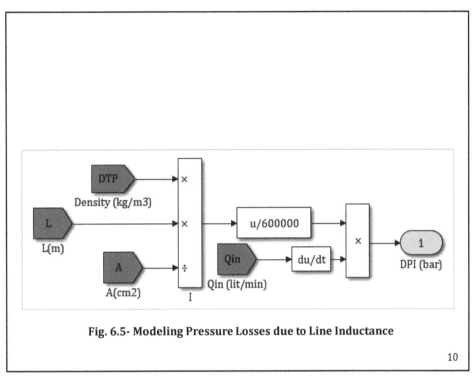

Fig. 6.5- Modeling Pressure Losses due to Line Inductance

10

10

6.1.3- Modeling Inlet Pressure of Transmission Line

$$P_{in} \, (bar) \; = \; P_{out} \, (bar) \; + \; \Delta P_R \, (bar) + \Delta P_I \, (bar) \qquad \textbf{6.9}$$

6.1.4- Modeling Outlet Flow due to Line Capacitance

Eq. 1.5D in Chapter 1 $\rightarrow$

$$\Delta P_{out} \; = \frac{\textbf{Equivalent Bulk Modulus } (\beta_E)}{\textbf{Container Volume (V)}} \int Q \, d \; \Rightarrow \frac{dP_{out}}{dt} = \frac{\beta_E}{V} \, [Q_{in} - Q_{out}]$$

$$\Rightarrow Q_{out} = Q_{in} - \frac{V}{\beta_E} \frac{dP_{out}}{dt} = Q_{in} - \frac{A \times L}{\beta_E} \frac{dP_{out}}{dt} \qquad \textbf{6.10A}$$

11

11

Equation 6.10B presents Eq. 6.10A with the units being balanced as follows:

$$Q_{out}\left[\frac{lit}{min}\right] = Q_{in}\left[\frac{lit}{min}\right] - \frac{A\,[cm^2] \times \frac{1}{10000} \times L[m]}{\beta_E\,[bar]} \frac{dP_{out}[bar]}{dt\,[s] \times \frac{1}{60}} \times 1000$$

$$Q_{out}\left[\frac{lit}{min}\right] = Q_{in}\left[\frac{lit}{min}\right] - \frac{6 \times A\,[cm^2] \times L[m]}{\beta_E\,[bar]} \frac{dP_{out}[bar]}{dt\,[s]} \qquad 6.10B$$

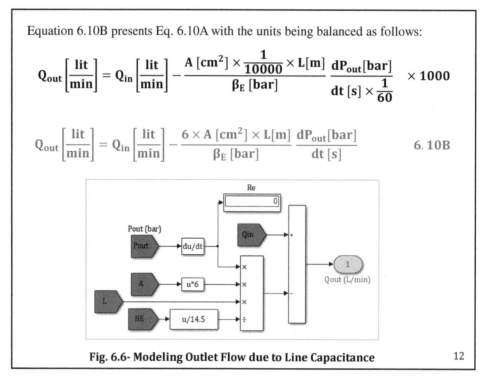

Fig. 6.6- Modeling Outlet Flow due to Line Capacitance   12

6.1.4- Case Studies for Modeling a Hydraulic Transmission Line

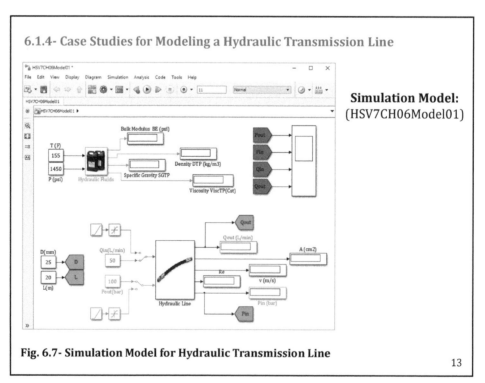

Simulation Model:
(HSV7CH06Model01)

Fig. 6.7- Simulation Model for Hydraulic Transmission Line   13

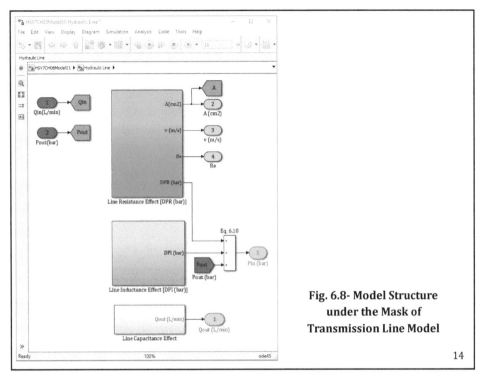

Fig. 6.8- Model Structure under the Mask of Transmission Line Model

14

14

6.1.4.1- Case Study 1: Steady State Condition

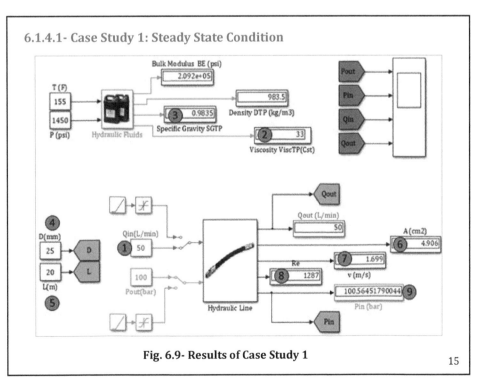

Fig. 6.9- Results of Case Study 1

15

15

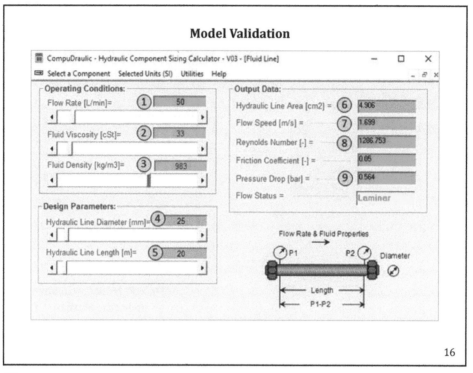

16

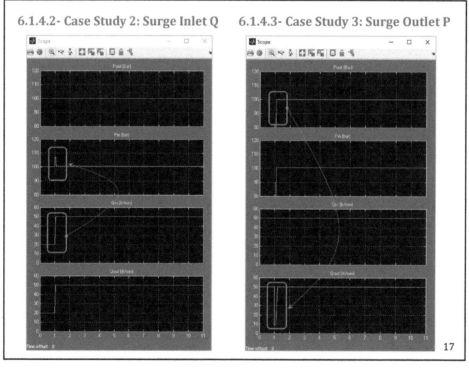

17

6.2- Modeling of Hydraulic Fittings

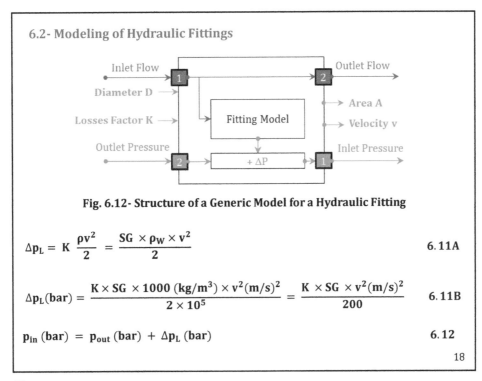

Fig. 6.12- Structure of a Generic Model for a Hydraulic Fitting

$$\Delta p_L = K \, \frac{\rho v^2}{2} = \frac{SG \times \rho_W \times v^2}{2} \qquad 6.11A$$

$$\Delta p_L (bar) = \frac{K \times SG \times 1000 \, (kg/m^3) \times v^2 (m/s)^2}{2 \times 10^5} = \frac{K \times SG \times v^2 (m/s)^2}{200} \qquad 6.11B$$

$$p_{in} \, (bar) = p_{out} \, (bar) + \Delta p_L \, (bar) \qquad 6.12$$

18

18

Where **K** is a lump sum resistivity factor known for every specific shape of fitting as

Fitting	90° bend	90° angle	T-Piece	Double angle	Valve
K	0.5 - 1	1.2	1.3	2	5 - 15
Sudden Contraction	Well-Rounded	Slightly Rounded	Sharp Edged	Shoulder	
K	0.04	0.2	0.5	0.8	
Sudden Expansion	Well-Rounded	Slightly Rounded	Sharp Edged	Shoulder	
K	1	1	1	1	

Fig. 6.13- Losses Factor in Fittings

19

19

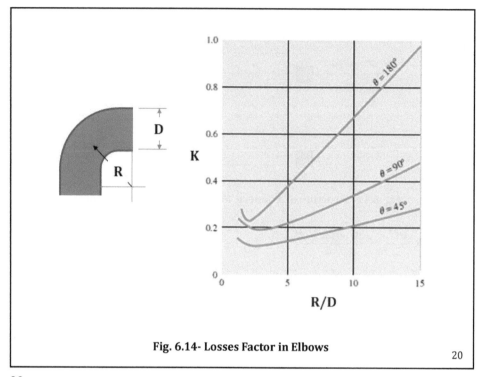

Fig. 6.14- Losses Factor in Elbows

20

20

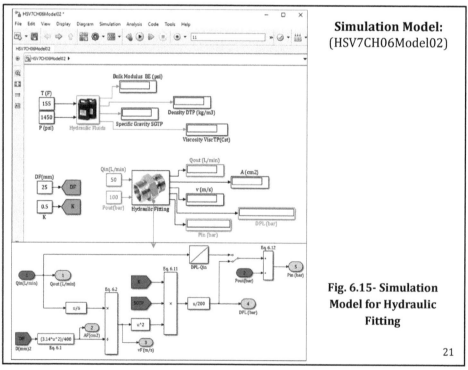

Simulation Model:
(HSV7CH06Model02)

Fig. 6.15- Simulation Model for Hydraulic Fitting

21

21

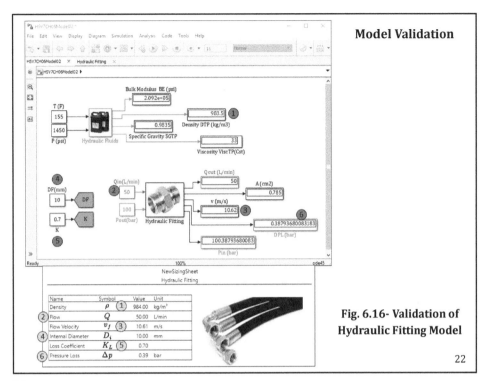

Model Validation

Fig. 6.16- Validation of Hydraulic Fitting Model

22

22

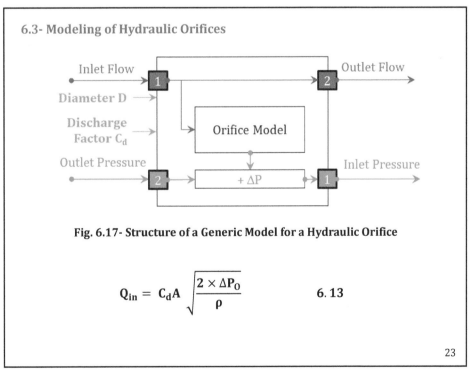

6.3- Modeling of Hydraulic Orifices

Fig. 6.17- Structure of a Generic Model for a Hydraulic Orifice

$$Q_{in} = C_d A \sqrt{\frac{2 \times \Delta P_0}{\rho}} \qquad 6.13$$

23

23

$$\Delta P_0 = \frac{\rho}{2}\left[\frac{Q_{in}}{C_d A}\right]^2 = \frac{SG \times \rho_w}{2}\left[\frac{Q_{in}}{C_d A}\right]^2 \qquad 6.14A$$

$$\Delta P_0(\text{Pascal}) = \frac{SG \times 1000\ (\text{kg/m}^3)}{2}\left[\frac{Q_{in}[\text{lit/min}] \times \frac{1}{60 \times 1000}}{C_d \times A\ [\text{mm}^2] \times \frac{1}{10^6}}\right]^2$$

$$\Delta P_0(\text{Pascal}) = \frac{SG \times 1000 \times [Q_{in}(\text{lit/min})]^2}{2 \times C_d^2 \times A^2\ [\text{mm}^2]}\left[\frac{100}{6}\right]^2\left[\frac{\text{kg.\,m}^6}{\text{m}^3}\frac{1}{\text{s}^2}\frac{1}{\text{m}^4}\right]$$

$$\Delta P_0(\text{Pascal}) = \frac{10^7 \times SG \times [Q_{in}(\text{lit/min})]^2}{72 \times C_d^2 \times A^2\ [\text{mm}^2]}\left[\frac{\text{kg.\,m}}{\text{s}^2}\frac{1}{\text{m}^2} = \frac{N}{\text{m}^2}\right]$$

$$\Delta P_0(\text{bar}) = \frac{\Delta P_0(\text{Pascal})}{10^5} = \frac{1000 \times SG \times [Q_{in}(\text{lit/min})]^2}{720 \times C_d^2 \times A^2\ [\text{mm}^2]} \qquad 6.14B$$

$$P_{in}\ (\text{bar}) = P_{out}\ (\text{bar}) + \Delta P_0\ (\text{bar}) \qquad 6.15$$

24

24

Other Modeling Method: An alternative easy method (lookup table).
Simulation Model: (HSV7CH06Model03)

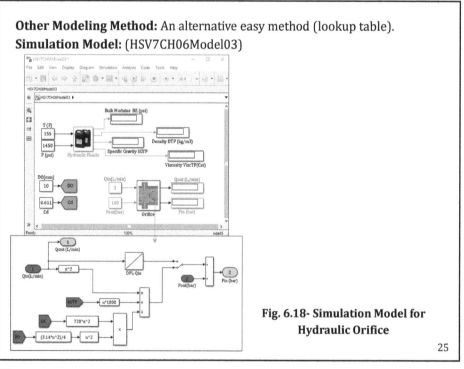

**Fig. 6.18- Simulation Model for
Hydraulic Orifice**

25

25

Model Validation

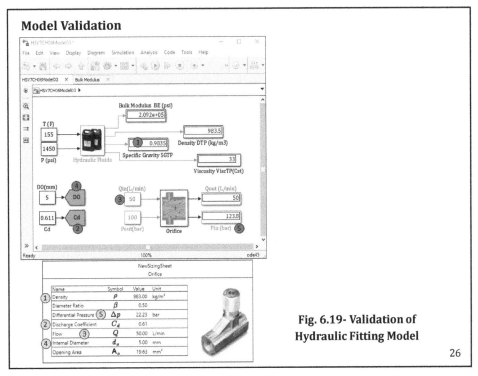

Fig. 6.19- Validation of Hydraulic Fitting Model

26

26

6.4- Modeling Hydraulic Transmission Line Assembly

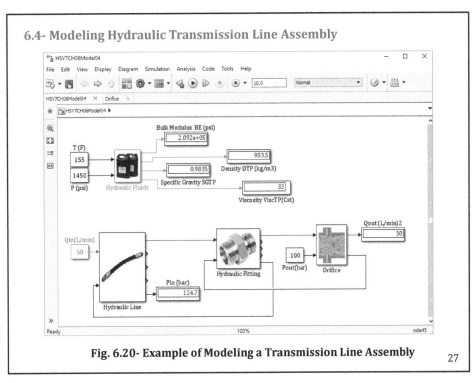

Fig. 6.20- Example of Modeling a Transmission Line Assembly

27

27

Chapter 6 Reviews

1. Lumped resistance of a hydraulic transmission line is related to?
   A. Pressure drop across the line due to fluid-fluid and fluid-wall friction.
   B. Pressure drop across the line due to fluid inertia.
   C. Pressure change inside a transmission line due to difference between inlet and outlet flow.
   D. None of the above.

2. Lumped inductance of a hydraulic transmission line is related to?
   A. Pressure drop across the line due to fluid-fluid and fluid-wall friction.
   B. Pressure drop across the line due to fluid inertia.
   C. Pressure change inside a transmission line due to difference between inlet and outlet flow.
   D. None of the above.

3. Lumped capacitance of a hydraulic transmission line is related to?
   A. Pressure drop across the line due to fluid-fluid and fluid-wall friction.
   B. Pressure drop across the line due to fluid inertia.
   C. Pressure change inside a transmission line due to difference between inlet and outlet flow.
   D. None of the above.

4. Reynold's Number above which the flow is definitely turbulent equal?
   A. 1000.
   B. 2000.
   C. 3000.
   D. 4000.

5. Roughness of a transmission line inside surface does not affect the line losses if?
   A. Fluid flow is laminar
   B. Fluid flow is transition
   C. Fluid flow is turbulent
   D. Fluid flows at low pressure.

Chapter 6 Assignment

Student Name: --- Student ID: ------------------

Date: --- Score: -----------------------

Assignment: use model (HSV7CH06Model03) to find out the overall pressure drop in the line assemble given:

- Line Diameter = 30 mm
- Line Length = 50 m
- Fitting Diameter = 10 mm
- Fitting Resistivity Coefficient = 1
- Orifice Diameter = 7 mm
- Orifice Discharge Coefficient = 0.611
- All other built in parameters in the model stays the same

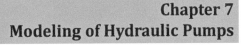

Chapter 7
Modeling of Hydraulic Pumps

Objectives:

This chapter presents the lumped modeling concept as applied for fixed and variable displacement pumps. This chapter considers situations where a pump works under a constant or variable pressure and driving speed. Models for pressure-compensated, displacement-controlled, and torque-limited pumps are developed.

0

0

Brief Contents:

1

1

7.1- Lumped Model Structure of a Unidirectional Hydraulic Pump

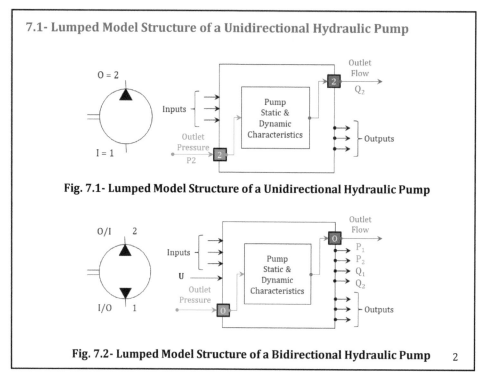

Fig. 7.1- Lumped Model Structure of a Unidirectional Hydraulic Pump

Fig. 7.2- Lumped Model Structure of a Bidirectional Hydraulic Pump 2

2

If **U** is positive, then port 1 is the intake port and port 2 is the delivery port
($P_1 = P_I$, $Q_1 = Q_I$, $P_2 = P_O$ and $Q_2 = Q_O$) and vise-versa.

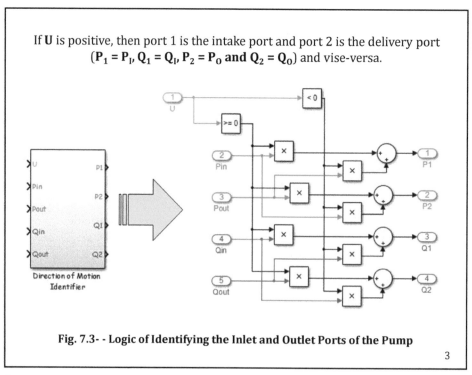

Fig. 7.3- - Logic of Identifying the Inlet and Outlet Ports of the Pump

3

3

7.3- Modeling Fixed Displacement Pumps

Pumps can work under fixed or variable driving speed and working pressure + Pump Characteristics reported in different forms
→ Various modeling scenarios

40 cc/rev pump size

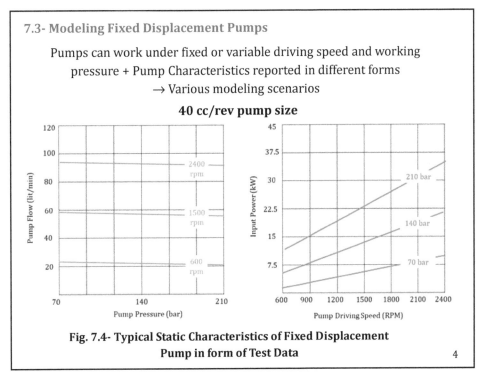

Fig. 7.4- Typical Static Characteristics of Fixed Displacement Pump in form of Test Data

4

4

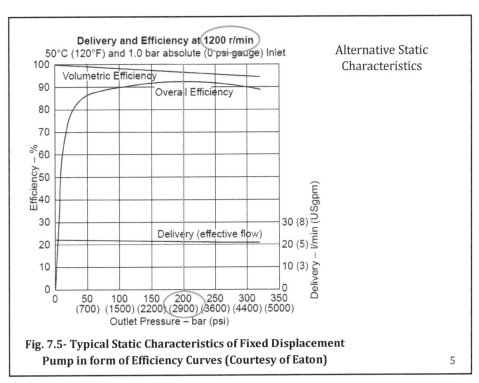

Alternative Static Characteristics

Fig. 7.5- Typical Static Characteristics of Fixed Displacement Pump in form of Efficiency Curves (Courtesy of Eaton)

5

5

7.4- Model #01 for an Ideal Fixed Displacement Pump
7.4.1- Model Features and Assumptions

Ideal Pump: (100% efficient)

Fig. 7.6- Pump Theoretical Flow Characteristics

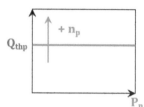

7.4.2- Pump Theoretical Flow Rate

$$Q_{thp} = D_p \times n_p \qquad\qquad 7.1$$

$$Q_{thp}(gpm) = \frac{D_p \left(\frac{in^3}{rev}\right) \times n_p(rpm)}{231} \qquad\qquad 7.1A$$

$$Q_{thp}(lit/min) = \frac{D_p \left(\frac{cc}{rev}\right) \times n_p(rpm)}{1000} \qquad\qquad 7.1B$$

6

6

7.4.3- Theoretical Torque Acting on the Pump Drive Shaft

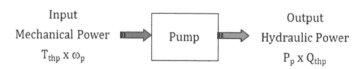

Input
Mechanical Power
$T_{thp} \times \omega_p$

Pump

Output
Hydraulic Power
$P_p \times Q_{thp}$

Fig. 7.7- Power Balance in an Ideal Pump

$$T_{thp} \times \omega_p = P_p \times Q_{thp}$$

$$\rightarrow \ T_{thp} \times 2\pi n_p = P_p \times D_p \times n_p \ \rightarrow T_{thp} = \frac{P_p \times D_p}{2\pi} \qquad 7.2$$

$$T_{thp}(lb.\,in) = \frac{P_p\,(psi) \times D_p(in^3/rev)}{2\pi} \qquad\qquad 7.2A$$

$$T_{thp}(N.\,m) = \frac{P_p\,(bar) \times D_p(cc/rev)}{20\pi} \qquad\qquad 7.2B$$

7

7

140

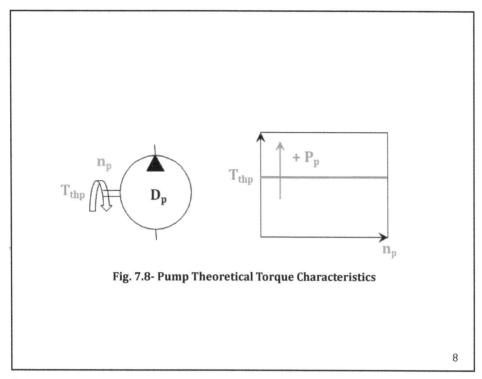

Fig. 7.8- Pump Theoretical Torque Characteristics

8

8

7.4.4- Model Structure

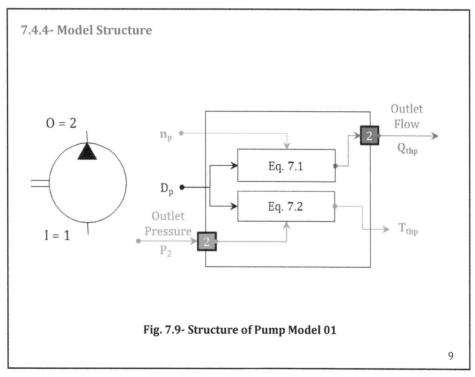

Fig. 7.9- Structure of Pump Model 01

9

9

7.4.5- Simulation Model

(HSV7CH07Model01)

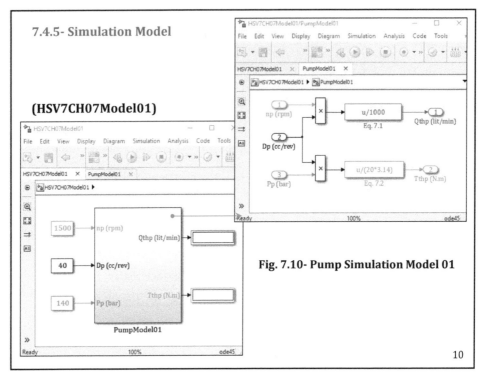

Fig. 7.10- Pump Simulation Model 01

10

10

7.5- Model #02A for a Fixed Displacement Pump Running at Constant Operating Conditions Based on Given Test Values

7.5.1- Model Features and Assumptions
- Practical Pump Working at Constant Operating Conditions.
- Given Test Data.

7.5.2- Pump Actual Flow and Volumetric E

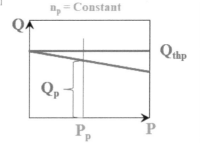

$$Q_p = e_{vp} \times Q_{thp} = Q_{thp} - q_L \qquad 7.3$$

Fig. 7.11- Q-p Characteristics of a Fixed Pump

11

11

7.5.3- Pump Input Power and Overall Efficiency

$$e_{op} = \frac{\text{Pump Output (Fluid)Power } [H_{out}]}{\text{Pump Input (Shaft)Power}[H_{in}]} = \frac{P_p(\text{psi}) \times Q_p(\text{gpm})}{1714 \times H_{in}(\text{HP})} \qquad 7.4A$$

$$e_{op} = \frac{\text{Pump Output (Fluid)Power } [H_{out}]}{\text{Pump Input (Shaft)Power}[H_{in}]} = \frac{P_p(\text{bar}) \times Q_p(\text{lit/min})}{600 \times H_{in}(\text{kW})} \qquad 7.4B$$

7.5.4- Pump Mechanical Efficiency

$$e_{mp} = e_{Op}/e_{vp} \qquad 7.5$$

12

12

7.5.5- Pump Actual Torque

$$T_p = \frac{T_{thp}}{e_{mp}} \qquad 7.6$$

Power losses H_L that is converted into heat in the pump.

$$H_L = H_{in} - H_{out} = H_{in} - H_{in} \times e_{Op} = H_{in}(1 - e_{Op}) \qquad 7.7$$

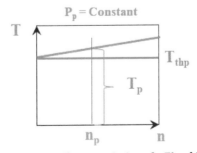

Fig. 7.12- T-n Characteristics of a Fixed Pump

13

13

143

7.5.6- Simulation Model

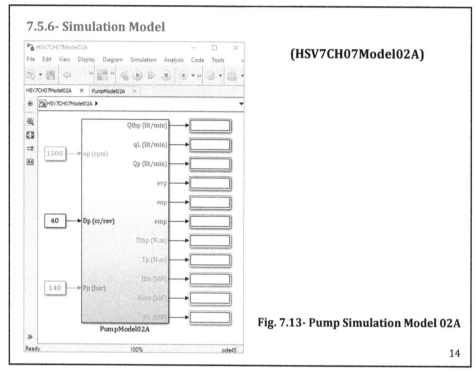

(HSV7CH07Model02A)

Fig. 7.13- Pump Simulation Model 02A

14

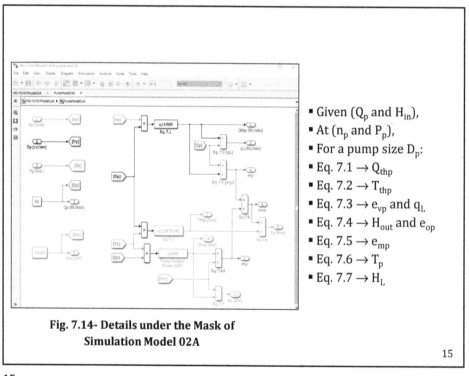

- Given (Q_p and H_{in}),
- At (n_p and P_p),
- For a pump size D_p:
- Eq. 7.1 → Q_{thp}
- Eq. 7.2 → T_{thp}
- Eq. 7.3 → e_{vp} and q_L
- Eq. 7.4 → H_{out} and e_{op}
- Eq. 7.5 → e_{mp}
- Eq. 7.6 → T_p
- Eq. 7.7 → H_L

**Fig. 7.14- Details under the Mask of
Simulation Model 02A**

15

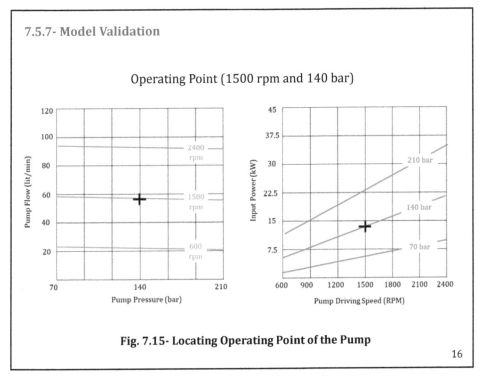

Fig. 7.15- Locating Operating Point of the Pump

16

16

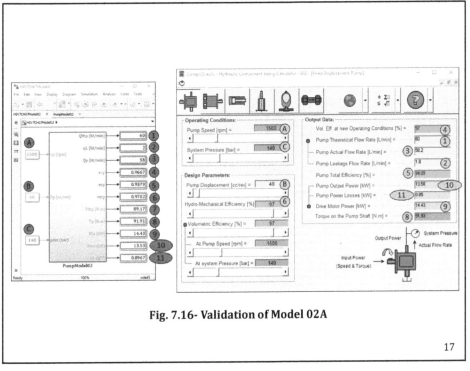

Fig. 7.16- Validation of Model 02A

17

17

7.6- Model #02B for a Fixed Displacement Pump Running at Constant Operating Conditions Based on Given Efficiency Values

7.6.1- Model Features and Assumptions
- Practical Pump Working at Constant Operating Conditions.
- Given Efficiency Curves.

7.6.2- Simulation Model

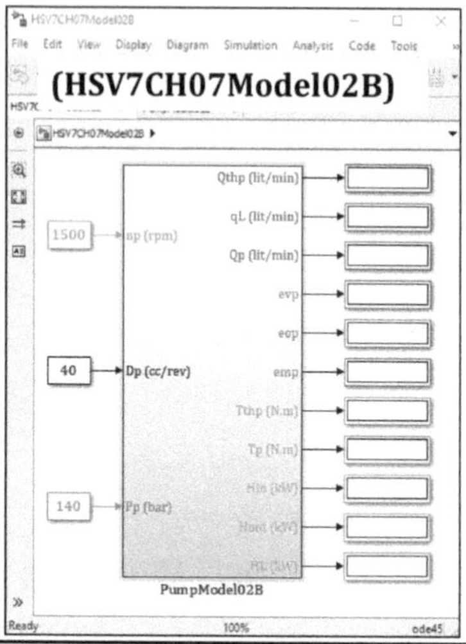

Fig. 7.17- Pump Simulation Model 02B

18

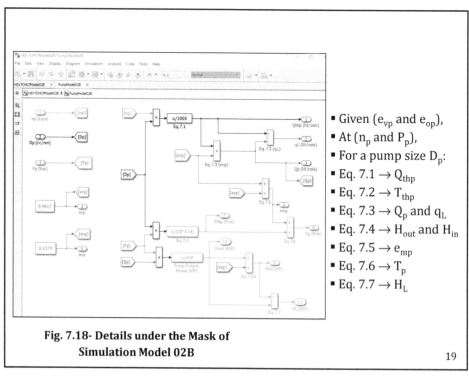

- Given (e_{vp} and e_{op}),
- At (n_p and P_p),
- For a pump size D_p:
- Eq. 7.1 $\rightarrow Q_{thp}$
- Eq. 7.2 $\rightarrow T_{thp}$
- Eq. 7.3 $\rightarrow Q_p$ and q_L
- Eq. 7.4 $\rightarrow H_{out}$ and H_{in}
- Eq. 7.5 $\rightarrow e_{mp}$
- Eq. 7.6 $\rightarrow T_p$
- Eq. 7.7 $\rightarrow H_L$

**Fig. 7.18- Details under the Mask of
Simulation Model 02B**

19

7.7- Model #03A for a Fixed Displacement Pump Running at Variable Operating Conditions Based on Given Test Data

7.7.1- Model Features and Assumptions

Given Test Data + Practical Pump Working at Variable Operating Conditions → Variable Efficiencies.

7.7.2- Mathematical Model

Test Data → two 2-D lookup tables

7.7.3- Simulation Model

Driving Speed n_p: Pump model receives n_p from the prime mover model.

Pump Outlet Pressure P_p: Pump model receives outlet pressure from the model of the component downstream the pump.

Pump Displacement D_p: Given characteristics are tied to that specific size.

Dynamic Effect of P_p (at D_p and n_p are Constants):

- $P_p \rightarrow$ Internal leakage + modulus of elasticity of the driving shaft.

Dynamic Effect of n_p (at D_p and p_p are Constants):

- Filling coefficient + fluid hydrodynamic effect + inertia of rotating mass.

20

20

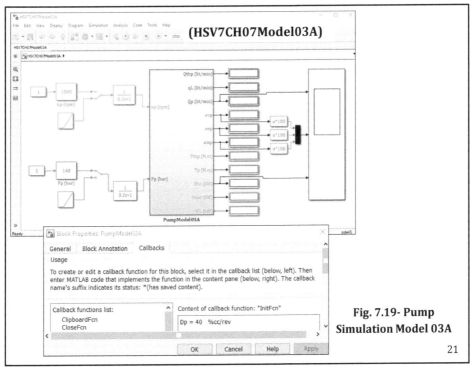

Fig. 7.19- Pump Simulation Model 03A

21

21

Model serves to simulate a pump under constant or variable conditions

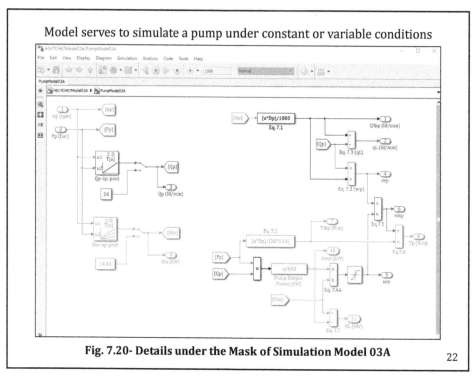

Fig. 7.20- Details under the Mask of Simulation Model 03A

22

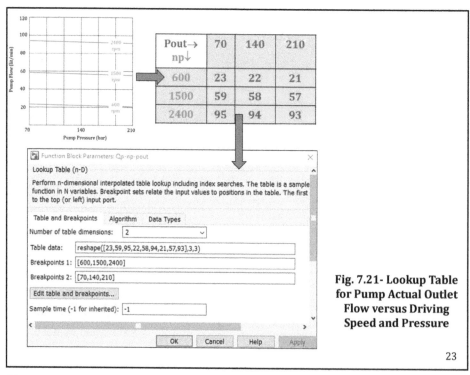

Fig. 7.21- Lookup Table for Pump Actual Outlet Flow versus Driving Speed and Pressure

23

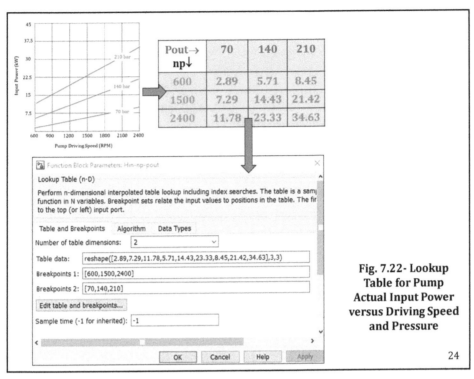

Fig. 7.22- Lookup Table for Pump Actual Input Power versus Driving Speed and Pressure

24

24

7.7.4- Model Validation

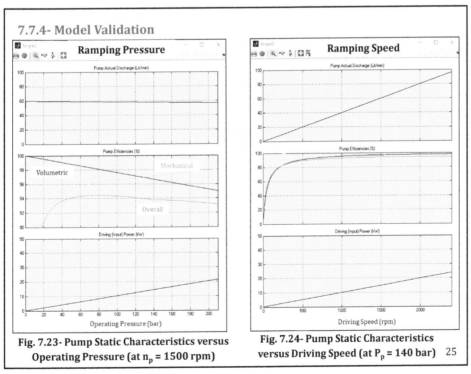

Fig. 7.23- Pump Static Characteristics versus Operating Pressure (at n_p = 1500 rpm)

Fig. 7.24- Pump Static Characteristics versus Driving Speed (at P_p = 140 bar) 25

25

7.8- Model #03B for a Fixed Displacement Pump Running at Variable Operating Conditions Based on Given Efficiency Curves

7.8.1- Model Features and Assumptions
Given Efficiency Curves + Practical Pump Working at Variable Operating Conditions → Variable Efficiencies.

7.8.2- Simulation Model

- Efficiency curves → two 2-D lookup tables.
- Same dynamic effect of speed and pressure change.

26

26

(HSV7CH07Model03B)

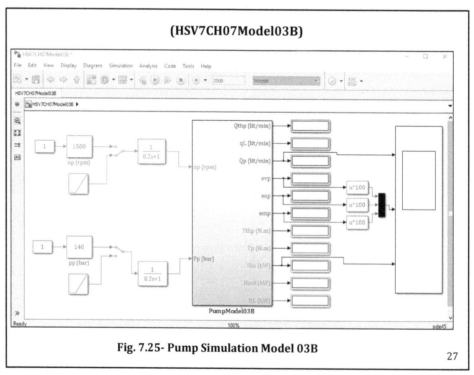

Fig. 7.25- Pump Simulation Model 03B

27

27

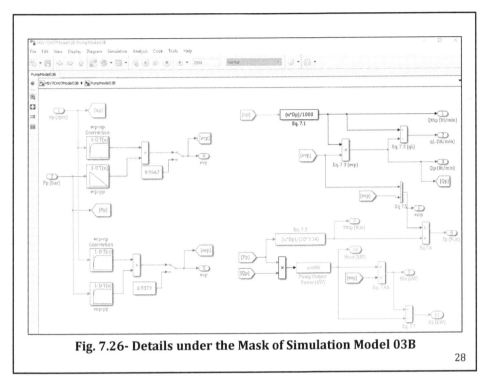

Fig. 7.26- Details under the Mask of Simulation Model 03B

28

28

7.8.3- Model Validation

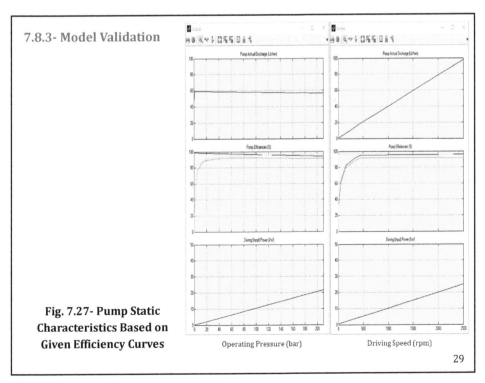

Operating Pressure (bar)   Driving Speed (rpm)

Fig. 7.27- Pump Static Characteristics Based on Given Efficiency Curves

29

29

7.9- Modeling Variable Displacement Pumps

Common control modes of variable displacement pumps as follows:

1. Pressure-Compensated pump.
2. Displacement-Controlled pump.
3. Torque-Limited (Constant-Power) pump.

- Other control modes (load-sense) out of scope of this textbook.
- If **Q-P** are the only available information → simple **Q-P** model is tied to a give pump size and speed.
- A model with full output requires full test data is available.

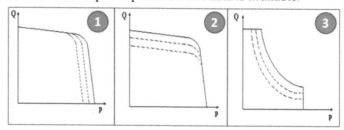

**Fig. 7.28- Common Control Modes of Variable Displacement Pump
(at Constant rpm)**

30

30

7.10- Lumped Modeling of Pressure-Compensated Pumps
7.10.1- Characteristics of Pressure-Compensated Pumps

A controller is actuated manually or remotely
(hydraulically by a pilot pressure or EH by an electrical variable signal).

- Modeled based on internal structure for pump development.
- Modeled based reported data for overall machine modeling.

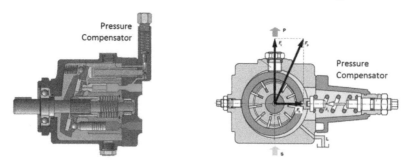

Fig. 7.29- Variable Displacement Pressure-Compensated Pumps

31

31

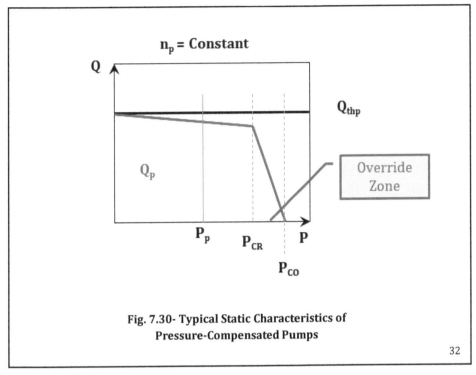

**Fig. 7.30- Typical Static Characteristics of
Pressure-Compensated Pumps**

32

Figure 7.31 shows typical test data for 56 cc/rev pressure-compensated
pump.

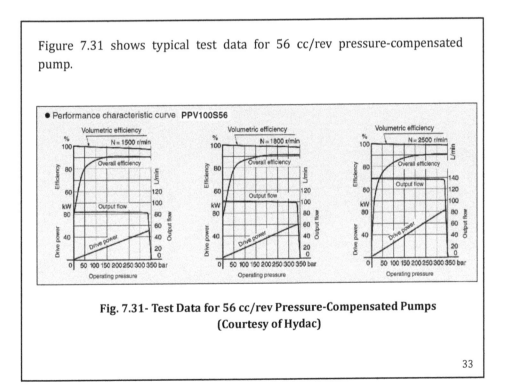

**Fig. 7.31- Test Data for 56 cc/rev Pressure-Compensated Pumps
(Courtesy of Hydac)**

33

7.10.2- Model #04A for Pressure-Compensated Pumps Based on Given Test Data

7.10.2.1- Model Features and Assumptions
- Consider the generic pump test data that are used in previous model.

7.10.2.2- Mathematical Model
- $P_p <= P_{CR} \rightarrow$ Same set of equations 7.1 through 7.7 are used.
- $P_p > P_{CR} \rightarrow$ Logic block was used to rout the calculation to the override zone.

7.10.2.3- Simulation Model
- **Pump Constants.:** values of cracking pressure, cut-off pressure, and pump maximum displacement are loaded to the block properties.
- **Dynamic Effect of Driving Speed Change (at D_p and P_p are Constants):** Same as in fixed displacement pumps.
- **Dynamic Effect of Operating Pressure Change (at n_p are Constants):**
- Ideally, we should simulate the dynamic effect before and after P_{CR}.
- However, for simplification, only dynamic of control unit is considered.
- Commonly, control units behave as a second order system.
- Reasonable natural frequency = 10 rad/s and damping ratio = 0.7.

34

34

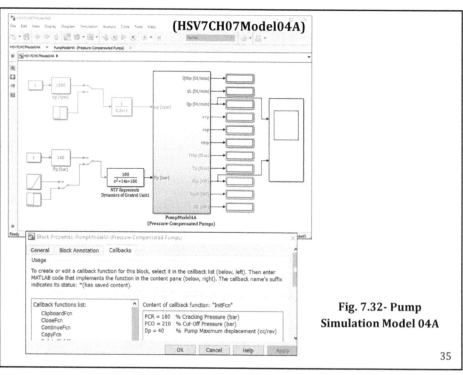

Fig. 7.32- Pump Simulation Model 04A

35

35

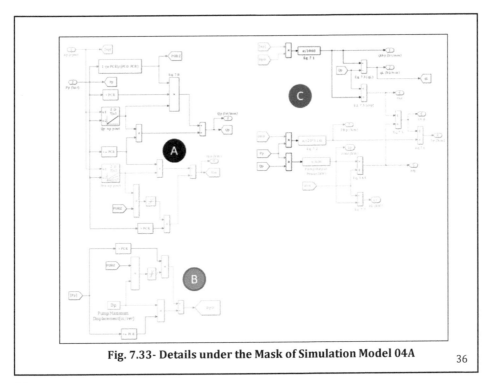

Fig. 7.33- Details under the Mask of Simulation Model 04A

36

36

Section A: Pump Flow and Input Power

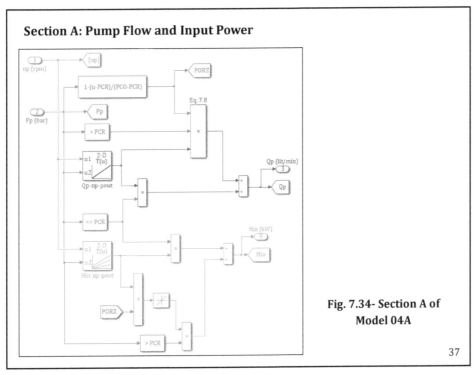

Fig. 7.34- Section A of Model 04A

37

37

Section B: Pump Displacement

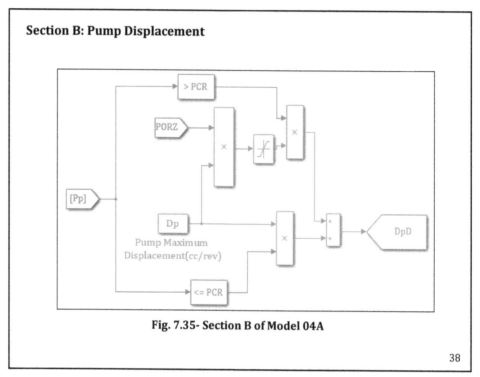

Fig. 7.35- Section B of Model 04A

38

Section C: Other Calculations

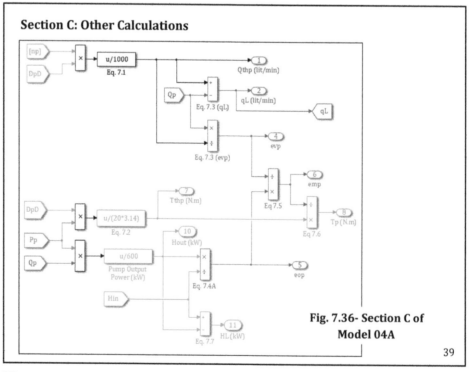

Fig. 7.36- Section C of Model 04A

39

7.10.2.4- Model Validation

Static Characteristics Validation:

- Pressure is ramped from zero to 210 bar (cracking pressure = 180 bar.

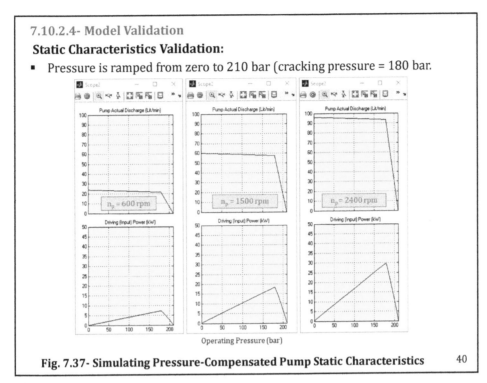

Fig. 7.37- Simulating Pressure-Compensated Pump Static Characteristics   40

40

Dynamic Characteristics Validation:

- Left side: effect of speed (at constant pressure of 140 bar).
- Right side: effect of pressure (at constant driving speed of 1500 rpm).

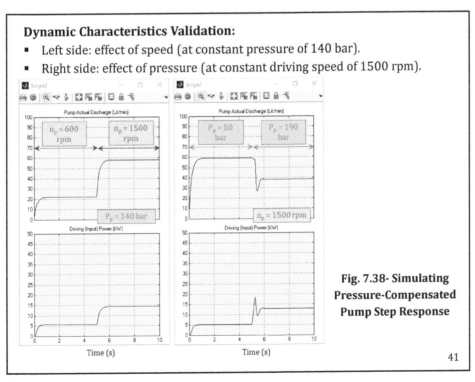

Fig. 7.38- Simulating Pressure-Compensated Pump Step Response

41

41

7.10.3- Model #04B for Pressure-Compensated Pumps Based on Given Efficiency Curves

7.10.3.1- Model Features and Assumptions
Given efficiency curves instead of pump test data.

7.10.2.2- Simulation Model

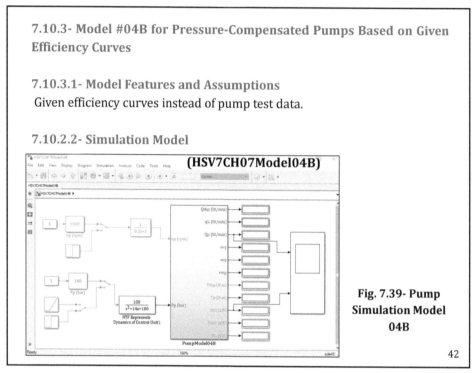

Fig. 7.39- Pump Simulation Model 04B

42

42

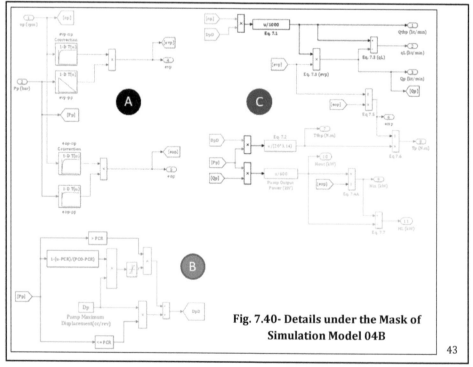

Fig. 7.40- Details under the Mask of Simulation Model 04B

43

43

7.10.4- Simplified Model #04C for Pressure-Compensated Pumps

Just based on given Q_p-P_p at one speed.

7.10.4.1- Case Study

Model 04C considers a pressure-compensated pump in the (UFPT).

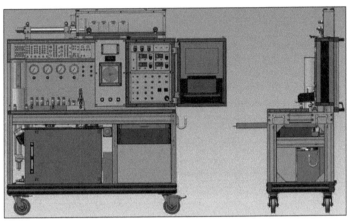

Fig. 7.41- Universal Fluid Power Trainer

44

44

7.10.4.2- Model Features and Assumptions

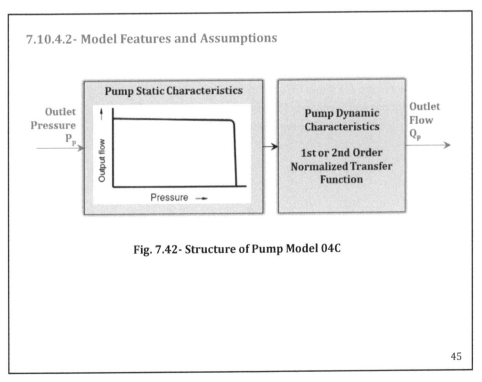

Fig. 7.42- Structure of Pump Model 04C

45

45

Pump Static Characteristics:

- Developed experimentally (ramping pressure).
- Loaded into a lookup table.

Ex.1 (Lab 9)

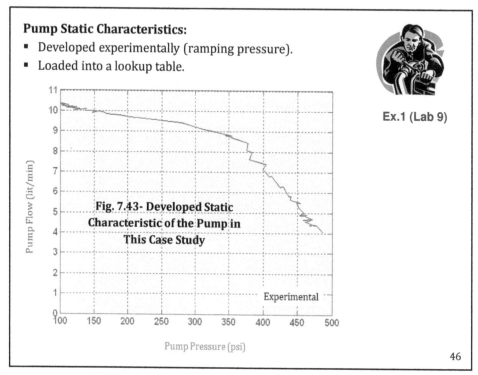

Fig. 7.43- Developed Static Characteristic of the Pump in This Case Study

46

46

Pump Dynamic Characteristics:

- Developed experimentally.
- Stepping down pressure.
- → D_p increased from min to maximum displacement in 1.5 seconds (no overshoot).
- → 1st Order NTF with time constant 300 ms.

Ex.2 (Lab 10)

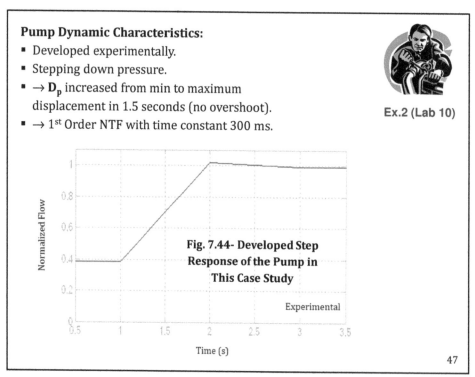

Fig. 7.44- Developed Step Response of the Pump in This Case Study

47

47

7.10.4.3- Simulation Model

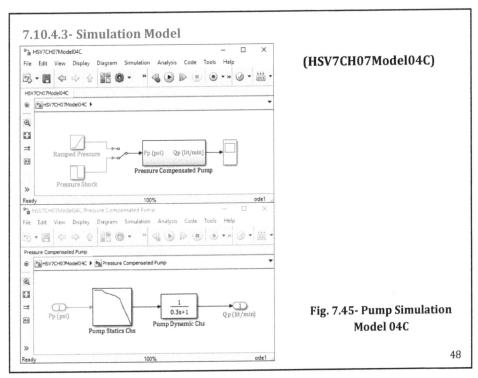

(HSV7CH07Model04C)

Fig. 7.45- Pump Simulation
Model 04C

48

48

7.10.4.4- Model Validation

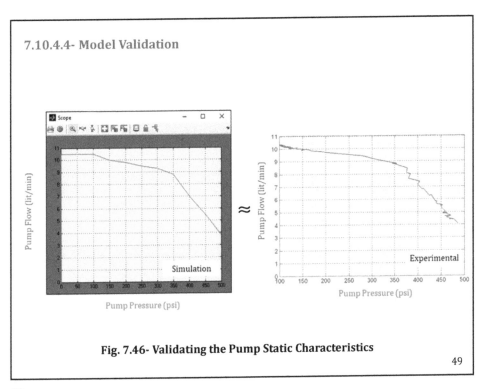

Fig. 7.46- Validating the Pump Static Characteristics

49

49

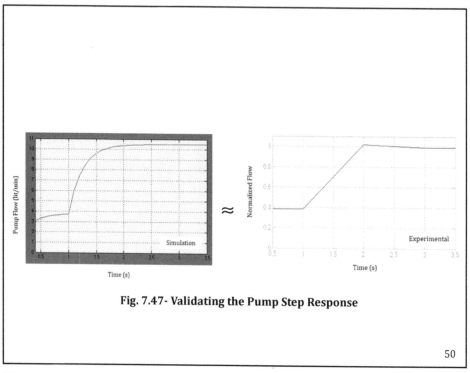

Fig. 7.47- Validating the Pump Step Response

50

50

7.11- Lumped Modeling of Displacement-Controlled Pumps
7.11.1- Characteristics of Displacement-Controlled Pumps

A controller is actuated manually or remotely (hydraulically by a pilot pressure or EH by an electrical variable signal).

- Modeled based on internal structure for pump development.
- Modeled based reported data for overall machine modeling.

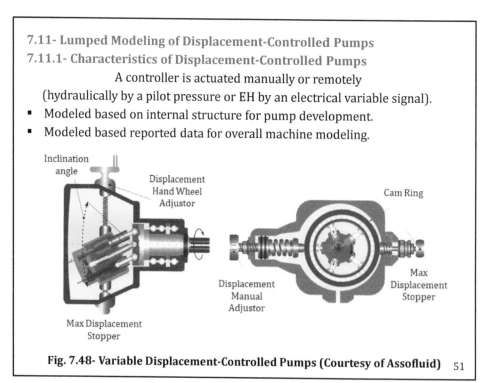

Fig. 7.48- Variable Displacement-Controlled Pumps (Courtesy of Assofluid) 51

51

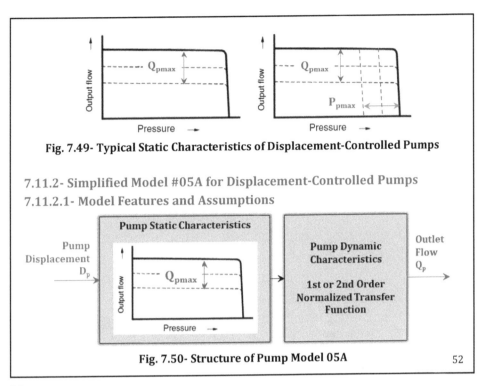

Fig. 7.49- Typical Static Characteristics of Displacement-Controlled Pumps

7.11.2- Simplified Model #05A for Displacement-Controlled Pumps

7.11.2.1- Model Features and Assumptions

Fig. 7.50- Structure of Pump Model 05A

52

7.11.2.2- Simulation Model

Static Characteristics: Q_p-P_p characteristics loaded into a 2-D lookup table.

Driving Speed: This model is valid for only a specific driving speed. Otherwise (e.g. mobile machines) 3-D lookup table is required.

Dynamic Characteristics: Second-order NTF represents the dynamics of the control unit. Damping ratio = 0.7 & natural frequency = 10 rad/sec.

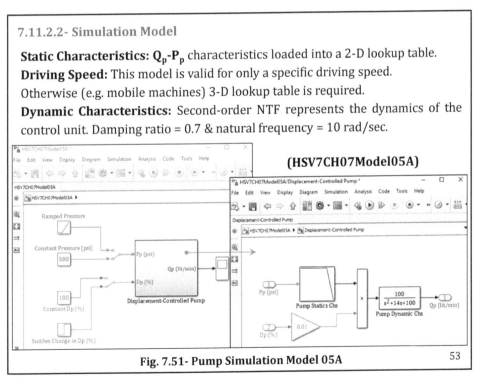

Fig. 7.51- Pump Simulation Model 05A

53

7.11.2.3- Model Validation

- **Upper part:** Pump flow varies.
- **Lower part:** step response.

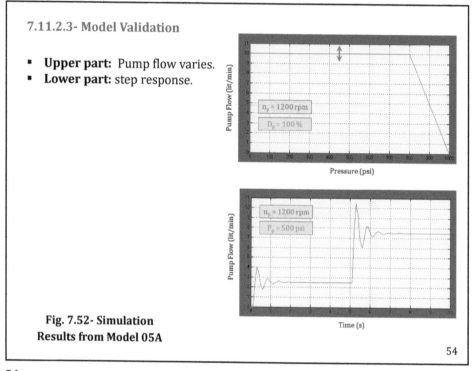

**Fig. 7.52- Simulation
Results from Model 05A**

54

7.11.3- Model #05B for Displacement-Controlled Pumps Based on Given Efficiency Curves

7.11.3.1- Model Features and Assumptions

- Basically, modeling such pumps requires the pump displacement as a varying input to the model. The root model 03A is not possible to be used because that model structure is built based on given test data loaded into lookup tables and this test data is tied to a specific pump displacement.
- Therefore, model 03B that is built based on given efficiency curves will be used as a root to build model 05B.

Assumption: Ideally efficiency curves should be available for various pump displacements; typically, 25%, 50%, 75% and 100%. So, with varying pump displacement, the proper efficiency curve is used. However, for simplified modeling process, efficiencies can be assumed as invariant with the pump displacement.

55

7.11.3.2- Simulation Model

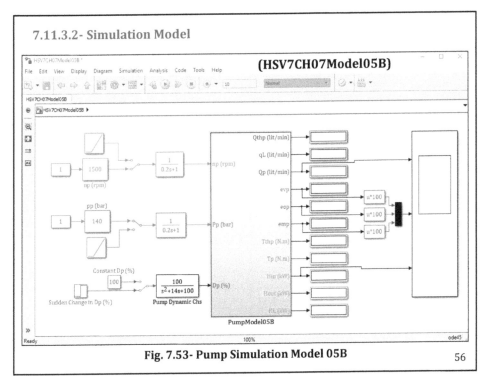

Fig. 7.53- Pump Simulation Model 05B

56

56

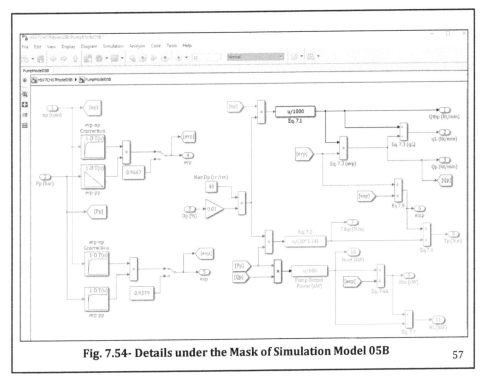

Fig. 7.54- Details under the Mask of Simulation Model 05B

57

57

7.11.3.3- Model Validation

Static Characteristics at Maximum Pump Displacement:

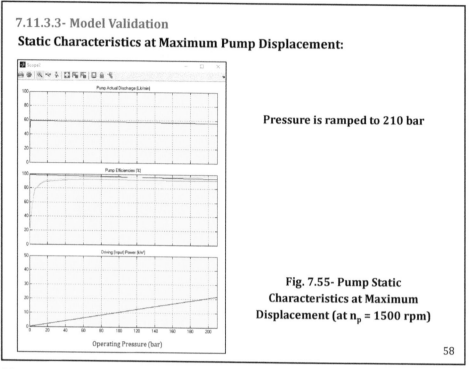

Pressure is ramped to 210 bar

Fig. 7.55- Pump Static Characteristics at Maximum Displacement (at n_p = 1500 rpm)

58

58

Static Characteristics at Reduced Pump Displacement:

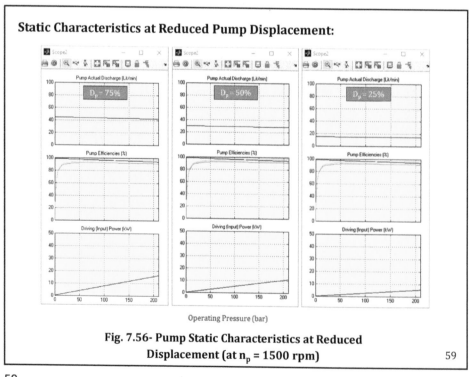

Fig. 7.56- Pump Static Characteristics at Reduced Displacement (at n_p = 1500 rpm)

59

59

Dynamic Characteristics:

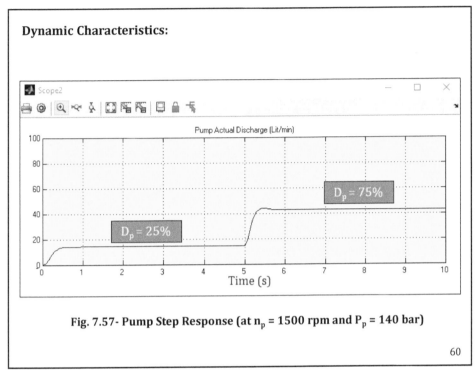

Fig. 7.57- Pump Step Response (at n_p = 1500 rpm and P_p = 140 bar)

60

60

7.12- Lumped Modeling of Torque-Limited Pumps
7.12.1- Characteristics of Torque-Limited Pumps

Torque Limited (Constant Power)

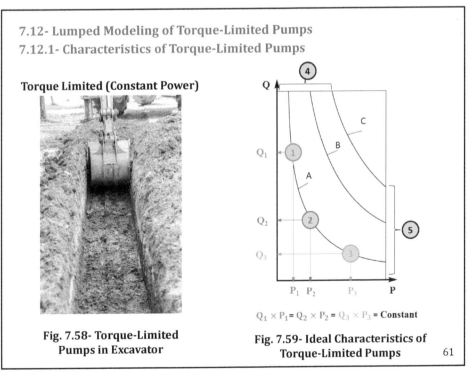

$$Q_1 \times P_1 = Q_2 \times P_2 = Q_3 \times P_3 = \text{Constant}$$

Fig. 7.58- Torque-Limited Pumps in Excavator

Fig. 7.59- Ideal Characteristics of Torque-Limited Pumps 61

61

7.12.1.1- Electro-Hydraulic Constant-Power Pumps

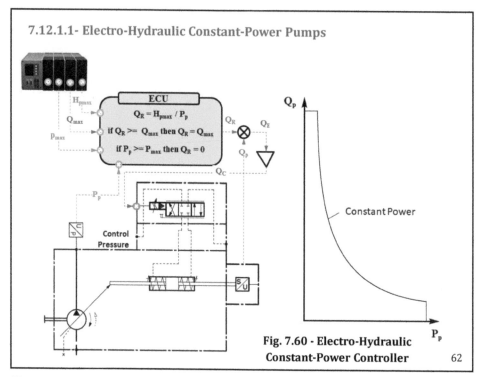

Fig. 7.60 - Electro-Hydraulic Constant-Power Controller 62

62

7.12.1.2- Hydro-Mechanical Constant-Power Pumps

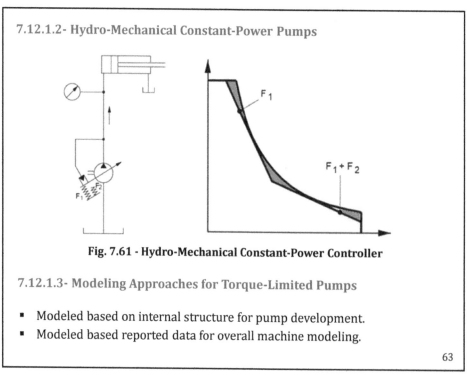

Fig. 7.61 - Hydro-Mechanical Constant-Power Controller

7.12.1.3- Modeling Approaches for Torque-Limited Pumps

- Modeled based on internal structure for pump development.
- Modeled based reported data for overall machine modeling.

63

7.12.2- Simplified Model #06A for Torque-Limited Pumps
7.12.2.1- Model Features and Assumptions

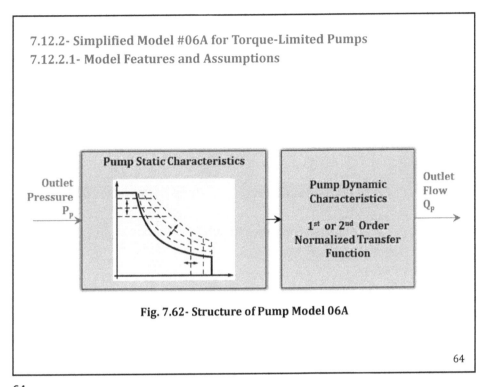

Fig. 7.62- Structure of Pump Model 06A

64

7.12.2.2- Simulation Model

Static Characteristics: Q_p-P_p characteristics loaded into a 2-D lookup table.

Driving Speed: This model is valid for only a specific driving speed.

Otherwise (e.g. mobile machines) 3-D lookup table is required.

Dynamic Characteristics: Second-order NTF represents the dynamics of the control unit. Damping ratio = 0.7 & natural frequency = 10 rad/sec.

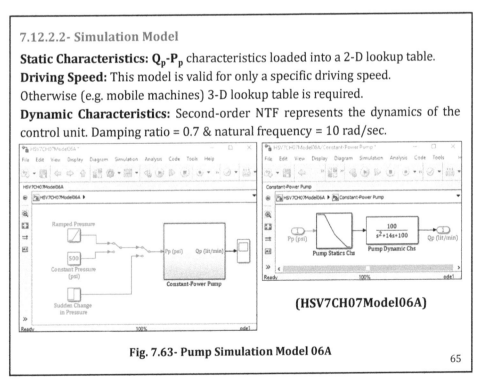

(HSV7CH07Model06A)

Fig. 7.63- Pump Simulation Model 06A

65

7.12.2.3- Model Validation

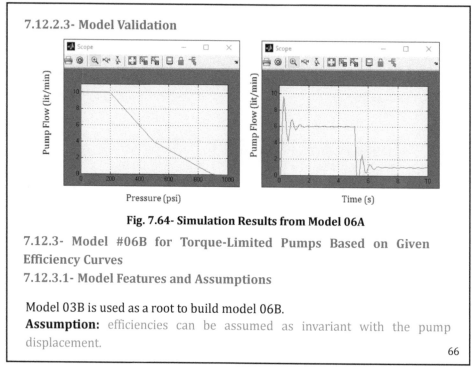

Fig. 7.64- Simulation Results from Model 06A

7.12.3- Model #06B for Torque-Limited Pumps Based on Given Efficiency Curves

7.12.3.1- Model Features and Assumptions

Model 03B is used as a root to build model 06B.

Assumption: efficiencies can be assumed as invariant with the pump displacement.

66

66

7.12.3.2- Simulation Model

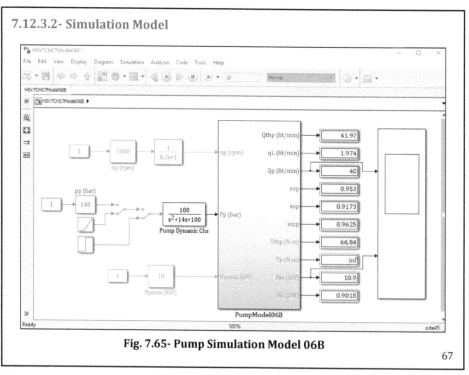

Fig. 7.65- Pump Simulation Model 06B

67

67

- Sections A: for efficiencies calculations
- Sections B: for power calculations
- Sections C: for flow calculations
- Sections D: for pump displacement calculations
- Sections E: for torque calculations

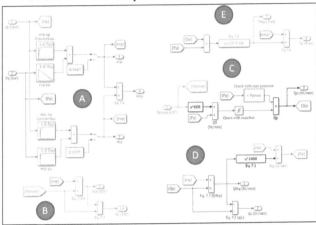

Fig. 7.66- Details under the Mask of Simulation Model 05B   68

68

7.12.3.3- Model Validation

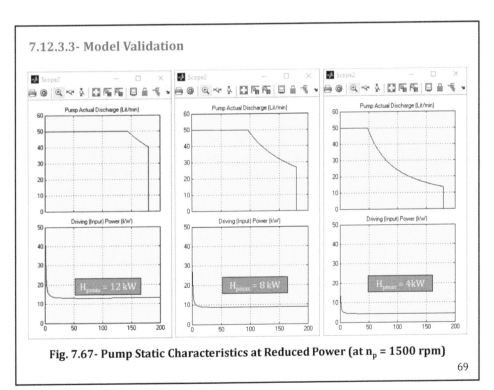

Fig. 7.67- Pump Static Characteristics at Reduced Power (at n_p = 1500 rpm)

69

69

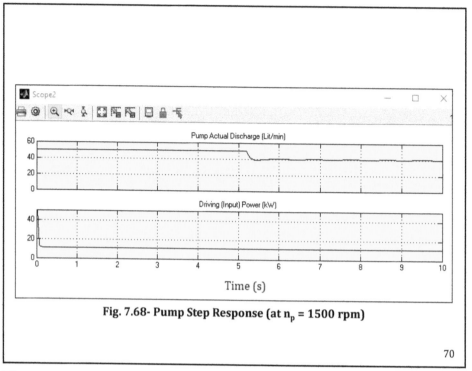

Fig. 7.68- Pump Step Response (at n$_p$ = 1500 rpm)

70

Chapter 7 Reviews

1. A lumped pump model that consider constant pump efficiencies is suitable for?
    A. An ideal pump.
    B. A pump that runs at constant speed under constant pressure.
    C. A pump that runs at variable speed under constant pressure.
    D. A pump that at constant rpm and variable pressure.

2. Internal leakage of a pump is represented by?
    A. Pump overall efficiency.
    B. Pump mechanical efficiency.
    C. Pump volumetric efficiency.
    D. None of the above

3. Internal friction of a pump is represented by?
    A. Pump overall efficiency.
    B. Pump mechanical efficiency.
    C. Pump volumetric efficiency.
    D. None of the above

4. Overall losses in the pump is represented by?
    A. Pump overall efficiency.
    B. Pump mechanical efficiency.
    C. Pump volumetric efficiency.
    D. None of the above

5. Theoretical driving torque on a pump shaft is function of?
    A. Pump displacement and mechanical efficiency.
    B. Pump pressure and mechanical efficiency.
    C. Pump volumetric and mechanical efficiency.
    D. Pump displacement and pump pressure

Chapter 7 Assignment

Student Name: --- Student ID: ------------------

Date: --- Score: -----------------------

Assignment: Use the model (HSV7CH07Model03B) to calculate the % increase of the driving torque (explain results) when:
1. Driving speed increased 50% on top of 1500 rpm at constant pressure of 140 bar.
2. Operating pressure increased 50% on top of 140 bar at constant speed of 1500 rpm.

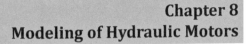

**Chapter 8
Modeling of Hydraulic Motors**

Objectives:

This chapter presents the lumped modeling concept as applied for fixed and variable displacement motors. This chapter considers situations where a motor works under a constant or variable torque and inlet flow. Models for two-position control, proportional control, and torque-limited motors are developed.

0

0

Brief Contents:

1

8.1- Lumped Model Structure of a Unidirectional Hydraulic Motor

- Incompressible flow $\rightarrow Q_0 = Q_{in} \rightarrow n_m$ & $\quad T_L \rightarrow \Delta P \rightarrow P_{in}$
- Model level based on available information, assumptions, and the typical applications.

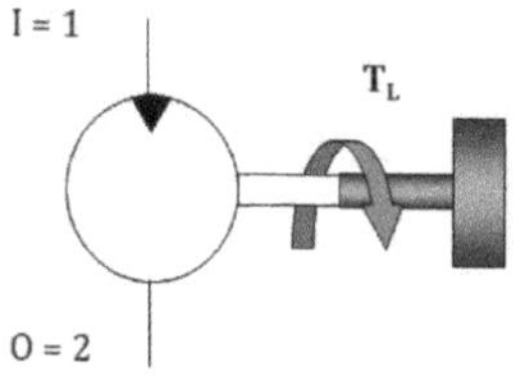

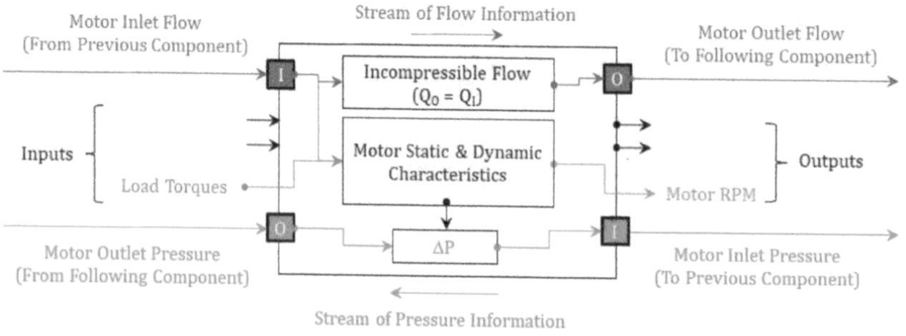

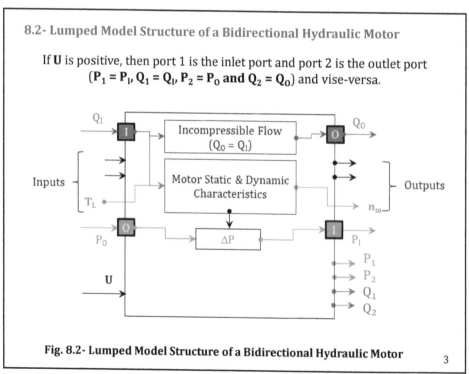

Fig. 8.1- Lumped Model Structure of a Unidirectional Hydraulic Motor 2

2

8.2- Lumped Model Structure of a Bidirectional Hydraulic Motor

If **U** is positive, then port 1 is the inlet port and port 2 is the outlet port
$(P_1 = P_I, Q_1 = Q_I, P_2 = P_0 \text{ and } Q_2 = Q_0)$ and vise-versa.

Fig. 8.2- Lumped Model Structure of a Bidirectional Hydraulic Motor 3

3

8.3- Modeling Fixed Displacement Motors

Motors can work under fixed or variable inlet flow and working torque
+ Motor Characteristics reported in different forms
→ Various modeling scenarios

- For motor sizing: input (T_m and n_m) and Output (Q_m & ΔP_m)
- For motor modeling: input (T_m and Q_m) and Output (ΔP_m & n_m)

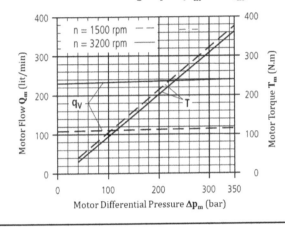

Fig. 8.3- Static Characteristics of a Fixed Displacement Motor in form of Test Data (Motor A4FM Series 71 – Courtesy of Bosch Rexroth)

4

4

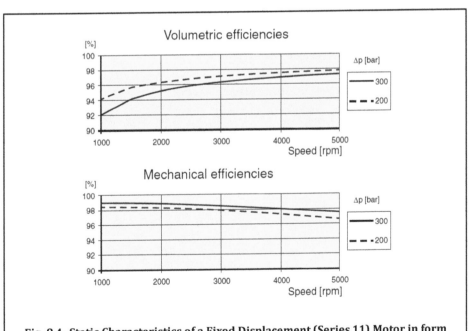

Fig. 8.4- Static Characteristics of a Fixed Displacement (Series 11) Motor in form of Efficiency Curves (Courtesy of Parker)

5

5

8.4- Model #01 for an Ideal Fixed Displacement Motor
8.4.1- Model Features and Assumptions
Ideal Motor: (100% efficient)

Fig. 8.5- Motor Theoretical RPM Characteristics

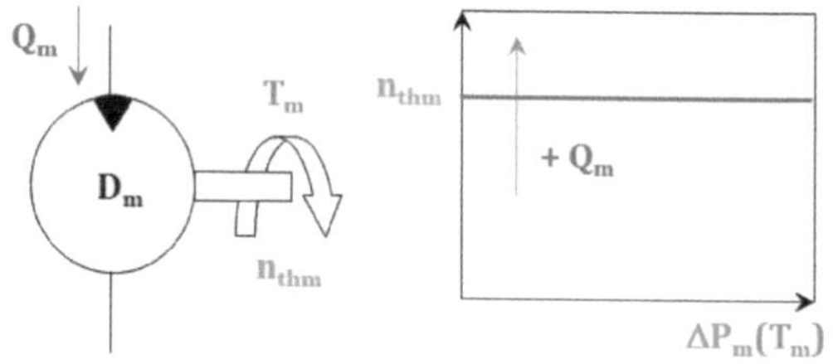

8.4.2- Motor Theoretical RPM

$$n_{thm} = Q_m / D_m \qquad\qquad 8.1$$

$$n_{thm}(rpm) = \frac{231 \times Q_m(gpm)}{D_m\left(\frac{in^3}{rev}\right)} \qquad\qquad 8.1A$$

$$n_{thm}(rpm) = \frac{1000 \times Q_m(lit/min)}{D_m\left(\frac{cc}{rev}\right)} \qquad\qquad 8.1B$$

6

6

8.4.3- Motor Torque Calculation
- **Based on the Source:** *Internal or External.*
- **Based on the Sign:** *Positive (Resistive) or Negative (Assistive).*
- **Based on the Motor Motion:** *Passive or Active.*

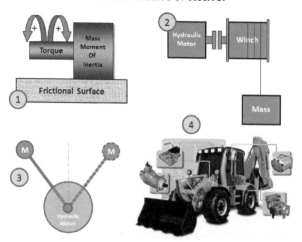

Fig. 8.6- Various Loading Conditions of Hydraulic Motors

7

7

$$T_V(N.m) = \omega \,(1/s) \times k_V\left(\frac{N.m}{1/s}\right) = \frac{2\pi n(rpm)}{60} \times k_V\left(\frac{N.m}{1/s}\right) \qquad 8.1C$$

$$T_I(N.m) = J\,(kg.m^2) \times \dot{\omega}\,(1/s^2) \qquad 8.1D$$

$$T_F(N.m) = F_N\,(N) \times R\,(m) \times k_F(-) \qquad 8.1E$$

$$T_m = T_L + T_V + T_I + T_F \qquad 8.1F$$

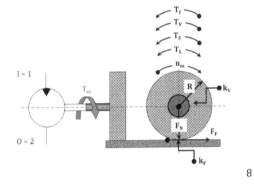

**Fig. 8.7- Hydraulic Motor
Torque Calculation**

8

8

8.4.4- Theoretical Differential Pressure Across the Motor

Input
Hydraulic Power
$\Delta P_{thm} \times Q_m$

Motor

Output
Mechanical Power
$T_m \times \omega_{thm}$

Fig. 8.8- Power Balance in an Ideal Motor

$$\Delta P_{thm} \times Q_m = T_m \times \omega_{thm}$$

$$\rightarrow \Delta P_{thm} \times D_m \times n_{thm} = T_m \times 2\pi n_{thm} \rightarrow \Delta P_{thm} = \frac{2\pi T_m}{D_m} \qquad 8.2$$

$$\Delta P_{thm}\,(psi) = \frac{2\pi \times T_m(lb.in)}{D_m(in^3/rev)} \qquad 8.2A$$

$$\Delta P_{thm}\,(bar) = \frac{20\pi \times T_m(N.m)}{D_m(cc/rev)} \qquad 8.2B$$

9

9

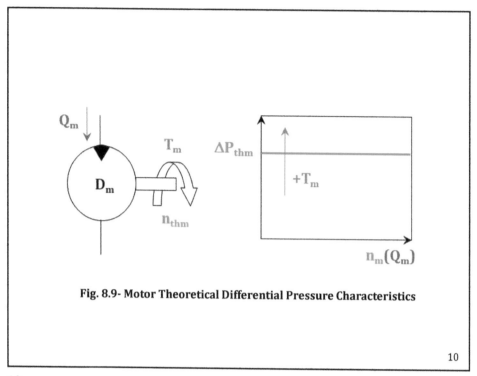

Fig. 8.9- Motor Theoretical Differential Pressure Characteristics

10

10

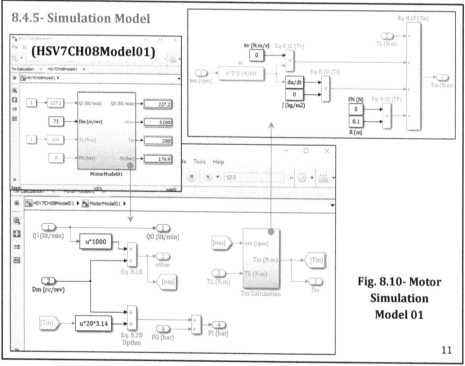

Fig. 8.10- Motor Simulation Model 01

11

11

8.4.6- Model Validation

- T_m = 200 N.m and n_m = 3200 →
- ΔP_{thm} (176.9 bar) < actual ΔP_m (200 bar) reported by the test data.
- The difference is due to internal friction.
- Q_{thm} (227.2 lit/min) < actual Q_m (240 lit/min) reported by the test data.
- The difference is due to motor internal leakage.

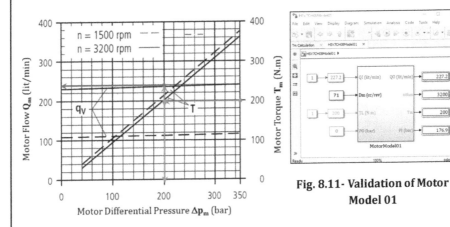

Fig. 8.11- Validation of Motor Model 01

12

12

8.5- Model #02A for a Fixed Displacement Motor Running at Constant Operating Conditions Based on Given Test Data

8.5.1- Model Features and Assumptions

- Practical Motor Working at Constant Operating Conditions.
- Given Test Data.

8.5.2- Motor Actual Speed and Volumetric Efficiency

- **On Curve A:** Q_{thm} → results in $n_m < n_{thm}$.
- **On Curve B:** $n_{thm,}$ → requires $Q_m > Q_{thm}$.

$$e_{vm} = Q_{thm}/Q_m = Q_{thm}/ (Q_{thm} + q_{Lm}) = n_m/n_{thm} \qquad 8.3$$

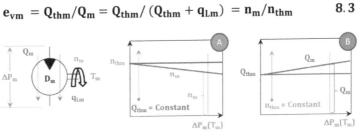

Fig. 8.12- Leakage Characteristics of a Fixed Motor 13

13

8.5.3- Motor Actual Torque and Mechanical Efficiency

- **On Curve A: $T_{thm} \rightarrow$ results in $\Delta P_m > \Delta P_{thm}$.**
- Δp_f is needed to overcome the motor internal friction.

- **On Curve B: $\Delta P_{hm} \rightarrow$ requires in $T_m < T_{thm}$**
- T_{Lm} must be compensated (subtracted from T_{thm}).

$$e_{mm} = T_m / T_{thm} = (T_{thm} + T_{Lm})/T_{thm} = \Delta P_{thm}/\Delta P_m \qquad 8.4$$

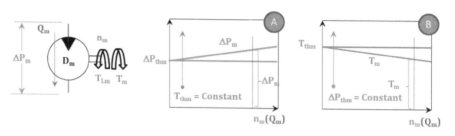

Fig. 8.13- Friction Characteristics of a Fixed Motor

14

14

8.5.4- Motor Overall Efficiency and Power

$$e_{om} = \frac{\text{Motor Output (Shaft)Power } [H_{outm}]}{\text{Motor Input (Fluid)Power} [H_{inm}]} = \frac{1714 \times H_{outm}(\text{Hp})}{\Delta P_m(\text{psi}) \times Q_m(\text{gpm})}$$

$$= e_{vm} \times e_{mm} \qquad 8.5A$$

$$e_{om} = \frac{\text{Motor Output (Shaft)Power} [H_{outm}]}{\text{Motor Input (Fluid)Power} [H_{inm}]} = \frac{600 \times H_{outm}(\text{kW})}{\Delta P_m(\text{bar}) \times Q_m(\text{lit/min})}$$

$$= e_{vm} \times e_{mm} \qquad 8.5B$$

$$H_{Lm} = H_{inm} - H_{outm} = H_{inm} - H_{inm} \times e_{Om} = H_{inm}(1 - e_{Om}) \, 8.6$$

15

15

8.5.5- Simulation Model

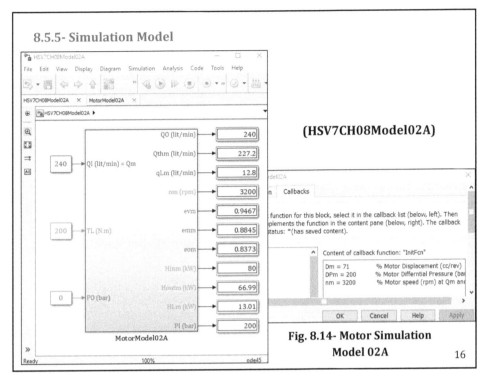

(HSV7CH08Model02A)

Fig. 8.14- Motor Simulation Model 02A

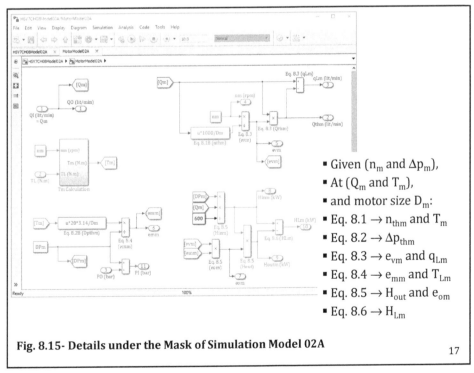

- Given (n_m and Δp_m),
- At (Q_m and T_m),
- and motor size D_m:
- Eq. 8.1 → n_{thm} and T_m
- Eq. 8.2 → Δp_{thm}
- Eq. 8.3 → e_{vm} and q_{Lm}
- Eq. 8.4 → e_{mm} and T_{Lm}
- Eq. 8.5 → H_{out} and e_{om}
- Eq. 8.6 → H_{Lm}

Fig. 8.15- Details under the Mask of Simulation Model 02A

17

8.5.6- Model Validation

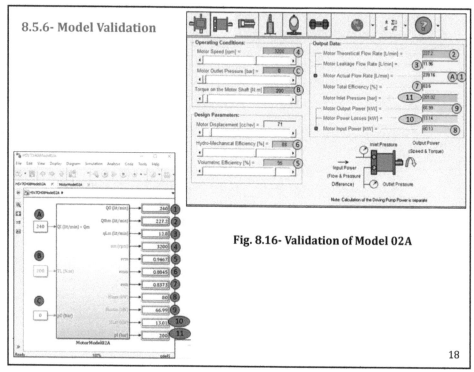

Fig. 8.16- Validation of Model 02A

18

8.6- Model #02B for a Fixed Displacement Motor Running at Constant Operating Conditions Based on Given Efficiency Values

8.6.1- Model Features and Assumptions

- Practical Motor Working at Constant Operating Conditions.
- Given Efficiency Curves.

8.6.2- Simulation Model

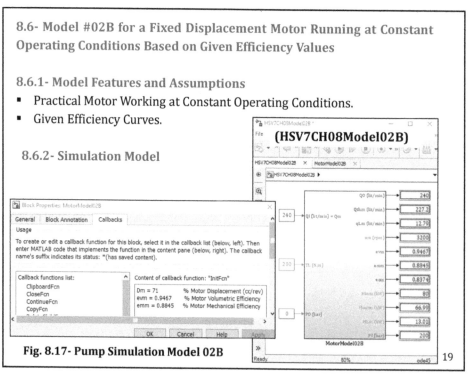

Fig. 8.17- Pump Simulation Model 02B

19

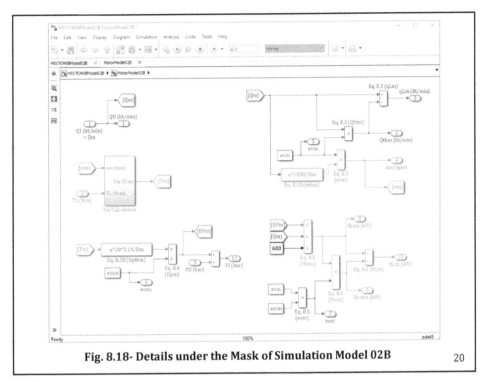

Fig. 8.18- Details under the Mask of Simulation Model 02B

20

20

8.7- Model #03A for a Fixed Displacement Motor Running at Variable Operating Conditions Based on Given Test Data

8.7.1- Model Features and Assumptions

Given Test Data + Practical Motor Working at Variable Operating Conditions
$\rightarrow$ Variable Efficiencies.

8.7.2- Mathematical Model

Test Data $\rightarrow$ two 2-D lookup tables

8.7.3- Simulation Model

Motor Inlet Flow Q_m: Motor model receives Q_m from the upstream model (a control valve or a variable displacement pump)

Motor Load Torque T_L: Motor model receives Q_m from load model.

Motor Displacement D_m: Given characteristics are tied to that specific size.

Dynamic Effect of Motor Torque (at D_m and Q_m are Constants):

- $\Delta P_m \rightarrow$ Internal leakage + modulus of elasticity of the driving shaft.

Dynamic Effect of Inlet Flow Change (at D_m and T_p are Constants):

- Filling coefficient + fluid hydrodynamic effect + inertia of rotating mass.

21

21

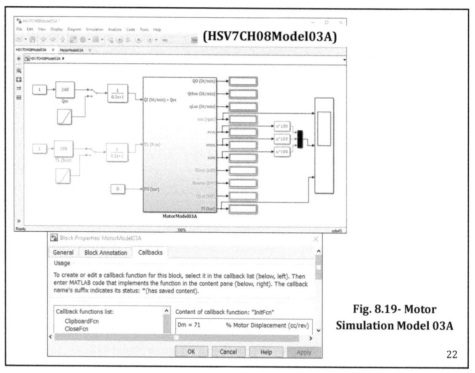

Fig. 8.19- Motor
Simulation Model 03A

22

22

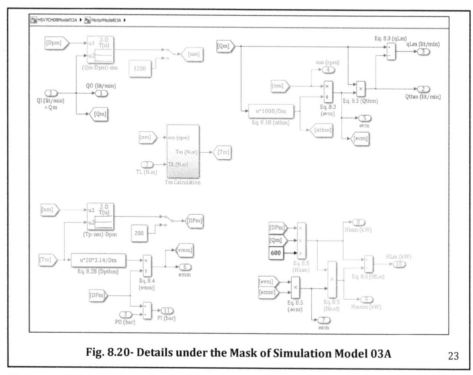

Fig. 8.20- Details under the Mask of Simulation Model 03A

23

23

Tm→ nthm↓	100	200	300
1500	90	190	290
2350	95	195	295
3200	100	200	300

Qm→ Dpm↓	100	200	300
100	1375	2750	4125
200	1325	2700	4075
300	1275	2650	4050

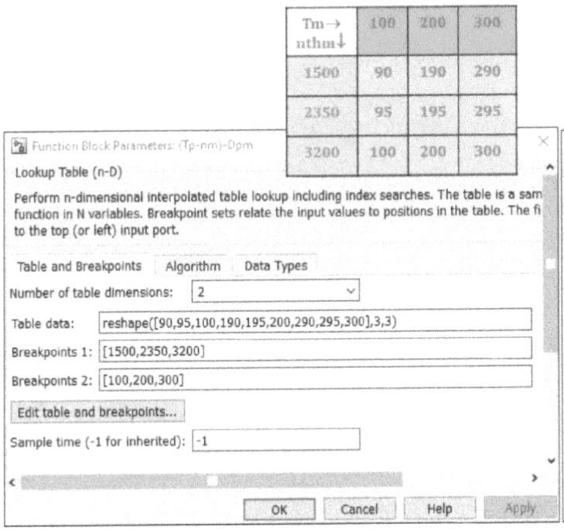

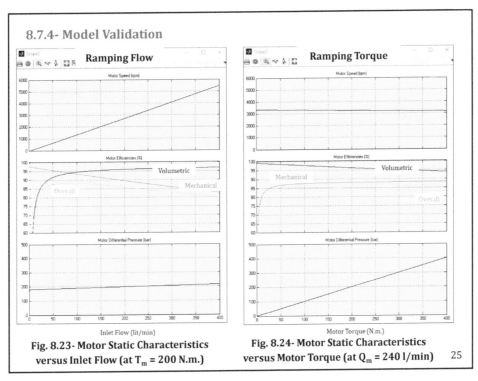

Fig. 8.21- Lookup Table for Motor Pressure-Torque Characteristics

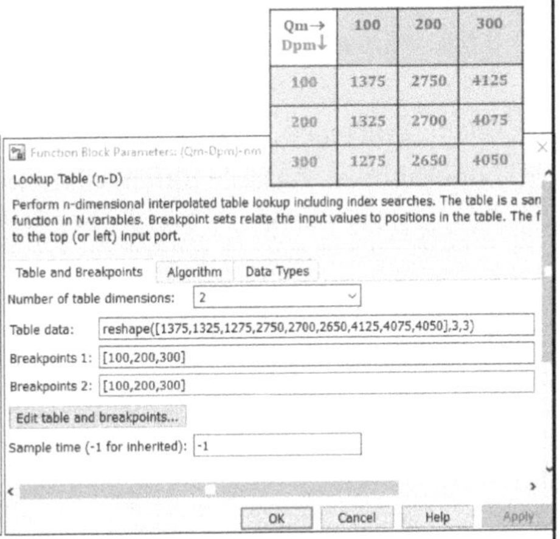

Fig. 8.22- Lookup Table for Motor Speed-Flow Characteristics

24

8.7.4- Model Validation

Fig. 8.23- Motor Static Characteristics versus Inlet Flow (at T_m = 200 N.m.)

Fig. 8.24- Motor Static Characteristics versus Motor Torque (at Q_m = 240 l/min)

25

8.8- Model #03B for a Fixed Displacement Motor Running at Variable Operating Conditions Based on Given Efficiency Curves

8.8.1- Model Features and Assumptions

Given Efficiency Curves + Practical Motor Working at Variable Operating Conditions → Variable Efficiencies.

8.8.2- Simulation Model
- Efficiency curves → two 2-D lookup tables.
- Same dynamic effect of torque and flow

26

26

(HSV7CH08Model03B)

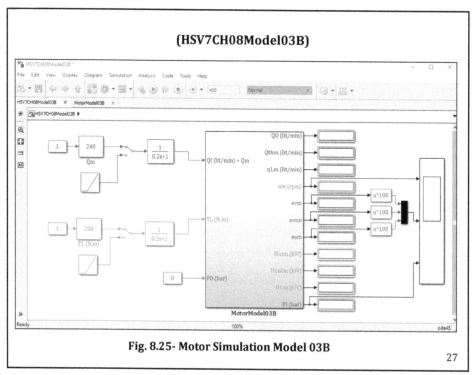

Fig. 8.25- Motor Simulation Model 03B

27

27

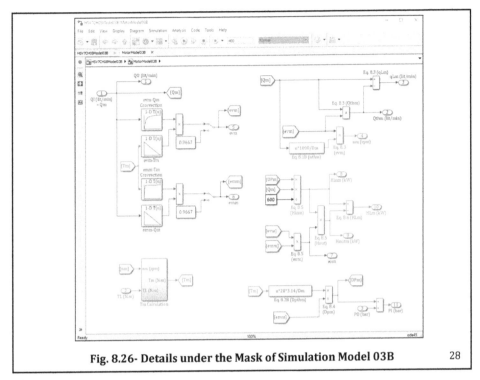

Fig. 8.26- Details under the Mask of Simulation Model 03B   28

28

8.8.3- Model Validation

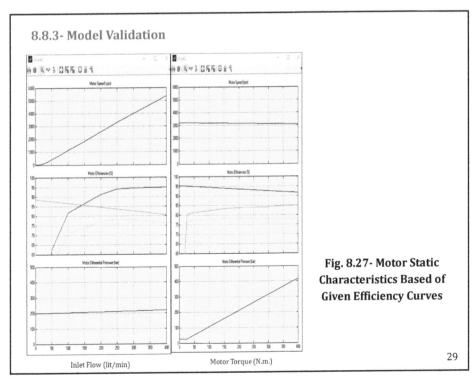

Inlet Flow (lit/min)   Motor Torque (N.m.)

Fig. 8.27- Motor Static Characteristics Based of Given Efficiency Curves

29

29

8.9- Modeling Variable Displacement Motors

A controller is actuated manually or remotely
(hydraulically by a pilot pressure or EH by an electrical variable signal).

- Modeled based on internal structure for pump development.
- Modeled based reported data for overall machine modeling.

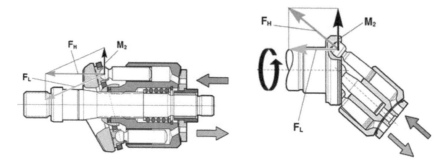

Fig. 8.28- Variable Displacement Motor Mechanisms
(Courtesy of Bosch Rexroth)

30

30

Control Modes:
1. Two Position Control.
2. Proportional Control.
3. Torque-Limiting (Constant-Power) Control

- If these characteristics are the only available information → simple model is tied to a motor size.
- A model with full output requires full test data is available.

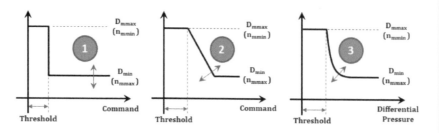

Fig. 8.29- Common Control Modes of Variable Displacement Motors

31

31

Min-Max Speed:
- Avoid running the motor erratically at speed lower than n_{min}.
- Avoid running the motor inefficiently at speed higher than n_{max}.

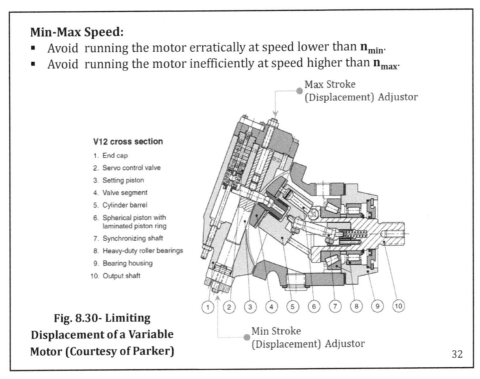

V12 cross section
1. End cap
2. Servo control valve
3. Setting piston
4. Valve segment
5. Cylinder barrel
6. Spherical piston with laminated piston ring
7. Synchronizing shaft
8. Heavy-duty roller bearings
9. Bearing housing
10. Output shaft

Max Stroke (Displacement) Adjustor

Min Stroke (Displacement) Adjustor

Fig. 8.30- Limiting Displacement of a Variable Motor (Courtesy of Parker)

32

32

8.10- Model #04 for Variable Displacement Motors
8.10.1- Model Features and Assumptions

- Easiest rout to model a variable displacement motor is based on given efficiency curves, assuming these efficiencies are valid for all range of motor displacements.

- Otherwise, multi-dimension lookup table should be used to determine efficiencies at various displacements.

33

33

8.10.2- Simulation Model

- motor displacement is commanded directly or could be pressure-related.
- A saturation block is used to limit the max-min displacement.

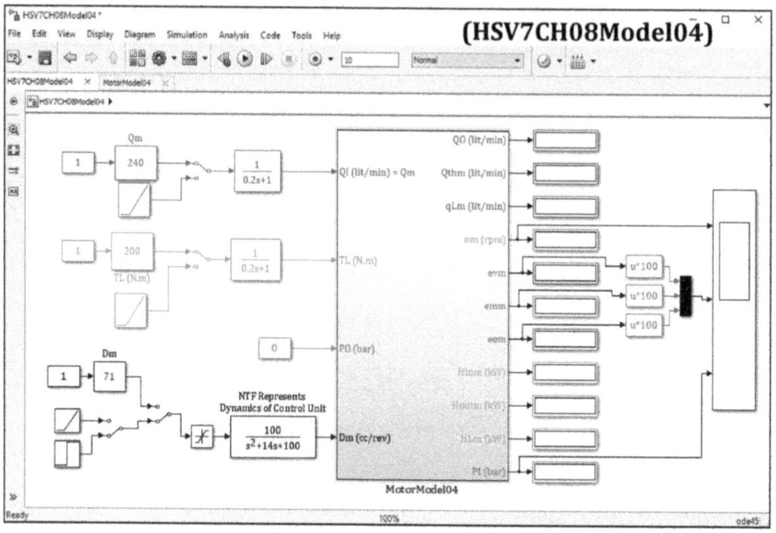

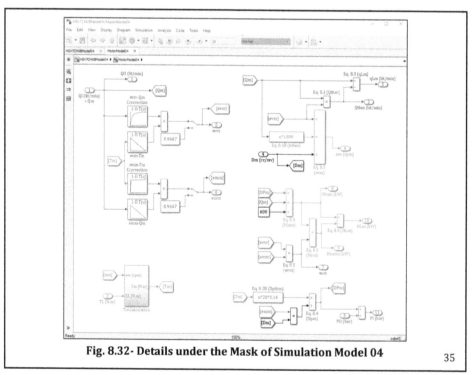

Fig. 8.31- Motor Simulation Model 04

34

Fig. 8.32- Details under the Mask of Simulation Model 04

35

8.10.3- Model Validation

Static Characteristics:

- Q_m = 240 lit/min
- T_m = 200 N.m.
- D_m is ramped up.
- D_m = limited between 40-70 cc/rev.
- Obviously, motor speed and differential pressure decrease with the increase of motor displacement.

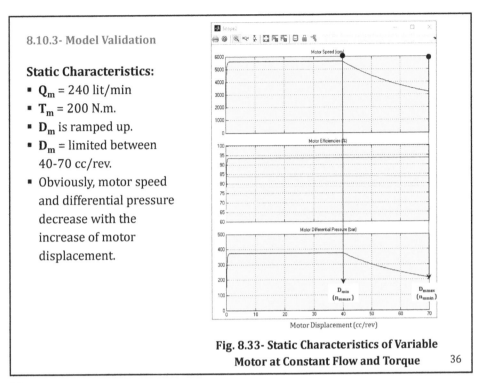

Fig. 8.33- Static Characteristics of Variable Motor at Constant Flow and Torque   36

36

Dynamic Characteristics:

- Q_m = 240 lit/min
- T_m = 200 N.m.
- D_m is stepped up from 40 to 60 cc/rev.
- Accordingly Speed and differential pressure decreased.

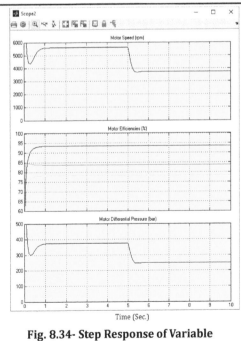

Fig. 8.34- Step Response of Variable Motor at Constant Flow and Torque   37

37

8.11- Simplified Model #05 for Valve-Controlled Fixed-Displacement Bidirectional Hydraulic Motor

8.11.1- Case Study
A servo valve-controlled hydraulic fixed-displacement motor in the (UFPT).

8.11.2- Model Features and Assumptions
Motor speed vs. input signal to a servo control valve.

Fig. 8.35- Structure of Motor Model 05

38

38

Motor Static Characteristics:
- Developed experimentally.
- Ramped input signal to the servo valve.
- Static characteristics are loaded into a lookup table.

Ex.3 (Lab 11)

Motor Dynamic Characteristics:
- Developed experimentally.
- Stepping up input command.
- $\rightarrow n_m$ increased from min to maximum with no overshoot.
- $\rightarrow$ 1st Order NTF with time constant 50 ms.

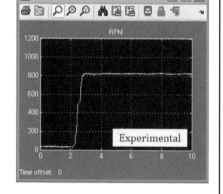

Ex.4 (Lab 12)   **Fig. 8.36- Developed Step Response of the Motor** 39

39

8.11.3- Simulation Model

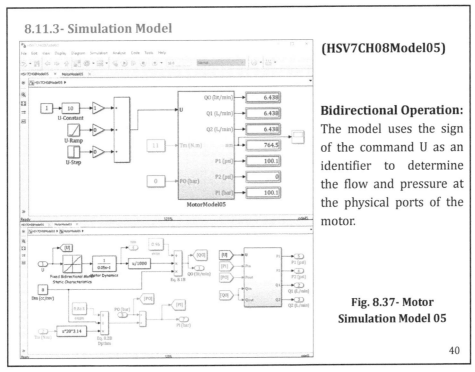

(HSV7CH08Model05)

Bidirectional Operation: The model uses the sign of the command U as an identifier to determine the flow and pressure at the physical ports of the motor.

Fig. 8.37- Motor Simulation Model 05

40

40

8.11.4- Model Validation

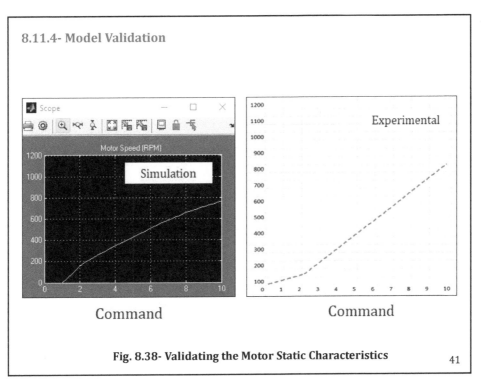

Fig. 8.38- Validating the Motor Static Characteristics

41

41

T_m = 11 (N.m) at n_m = 764 rpm

→ ΔP_m = 100 bar

→ Matches test data reported by the motor's manufacture.

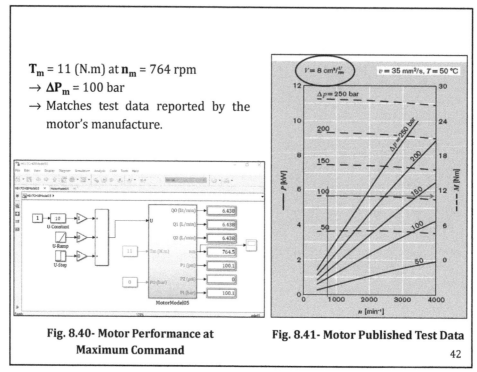

Fig. 8.40- Motor Performance at Maximum Command

Fig. 8.41- Motor Published Test Data

42

42

196

Chapter 8 Reviews

1. A hydraulic motor that has a displacement D_1, generates torque T_1 at a differential pressure 100 bar. If the motor displacement is doubled, then the new differential pressure is?
   A. 200 bar.
   B. 100 bar.
   C. 50 bar.
   D. Pressure dose not affected by the motor size

2. A hydraulic motor that has a displacement D_1, receives constant flow and running at 1000 rpm at a differential pressure 100 bar. If the motor displacement is doubled, then the new rpm is?
   A. 500 rpm
   B. 1500 rpm
   C. 2000 rpm
   D. Motor speed dose not affected by the motor size

3. When the internal friction of a hydraulic motor is reduced due to better lubrication, which of the following statement is the most correct?
   A. Motor mechanical efficiency is improved
   B. Motor overall efficiency is improved
   C. Motor torque increased
   D. All of the above is correct

4. When the internal leakage of a hydraulic motor is increased, which of the following statement is considered FALSE?
   A. Motor overheats
   B. Motor speed increased
   C. Motor requires more flow to sustain the same speed
   D. All of the above is correct

5. Which of the following statement is considered TRUE?
   A. Speed of a hydraulic motor does not affected by internal leakage
   B. Torque of a hydraulic motor does not affected by internal friction
   C. A variable speed motor can't be fully de-stroked, otherwise it will stall
   D. All of the above is true

Chapter 8 Assignment

Student Name: -- Student ID: ------------------

Date: --- Score: -----------------------

Assignment: Using model (HSV7CH08Model04), develop the dynamic response of a variable displacement motor when the displacement of the motor is reduced suddenly from 80 cc/rev to 40 cc/rev. Use 10 seconds time window with and motor displacement change occur at half of the tie window.

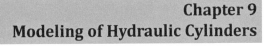

Chapter 9
Modeling of Hydraulic Cylinders

Objectives:

This chapter presents the lumped modeling concept as applied for double-acting hydraulic cylinders. This chapter considers situations where a cylinder works under a constant or variable external load and inlet flow. Calculations for cylinder slowing due to leakage, cylinder drift due to oil compressibility, and pressure increase due to thermal expansion are presented.

0

0

Brief Contents:

9.1- Lumped Model Structure of a Double-Acting Hydraulic Cylinder

9.2- Lumped Model #01 for Hydraulic Cylinder

9.3- Modeling Cylinder Drifting due to Oil Bulk Modulus

9.4- Modeling of Pressure Increase due to Thermal Expansion

1

1

9.1- Lumped Model Structure of a Double-Acting Hydraulic Cylinder

Modeling Approach:
Different configurations of hydraulic cylinders (mill-type & tie-rod)
Modeled based on internal structure for pump development.
Modeled based reported data for overall machine modeling.

Fig. 9.1- Hydraulic Cylinder Structural Configurations

2

2

- Incompressible flow → cylinder speed = f(inlet flow)
- External *load* F_L → differential pressure across the cylinder.
- Cylinder dimensions and dynamic characteristics are loaded to the moel properties.
- An identifier **U** is used to identify the cylinder direction of motion and, accordingly, working conditions at piston side (1) and rod side (2).

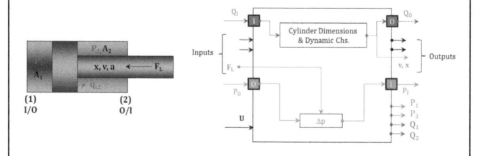

Fig. 9.2- Lumped Model Structure of a Double-Acting Hydraulic Cylinder

3

3

9.2- Lumped Model #01 for Hydraulic Cylinder
9.2.1- Mathematical Model
9.2.1.1- Cylinder Effective Areas

$$A_C(\text{in}^2) = \frac{\pi}{4} D_C^2 \, (\text{in}) \qquad\qquad \textbf{9.1A}$$

$$A_C(\text{cm}^2) = \frac{\pi}{4} D_C^2 \, (\text{mm}) \qquad\qquad \textbf{9.1B}$$

$$A_R(\text{in}^2) = \frac{\pi}{4} \left[D_C^2 \, (\text{in}) - D_R^2 \, (\text{in}) \right] \qquad\qquad \textbf{9.1C}$$

$$A_R(\text{cm}^2) = \frac{\pi}{4} \left[D_C^2 \, (\text{mm}) - D_R^2 \, (\text{mm}) \right] \qquad\qquad \textbf{9.1D}$$

4

4

9.2.1.2- Cylinder Inlet Pressure

By applying force balance on the cylinder piston,

$$P_I \times A_I - P_O \times A_O = F_C \;\rightarrow\; P_I = [F_C + P_O \times A_O]/A_I \qquad\qquad \textbf{9.2}$$

$$P_I \,(\text{psi}) = \frac{F_C(\text{lb}) + P_O(\text{psi}) \times A_O(\text{in}^2)}{A_I(\text{in}^2)} \qquad\qquad \textbf{9.2A}$$

$$P_I \,(\text{bar}) = \frac{0.1 \times F_C(\text{N}) + P_O(\text{bar}) \times A_O(\text{cm}^2)}{A_I(\text{cm}^2)} \qquad\qquad \textbf{9.2B}$$

5

5

9.2.1.3- Cylinder Internal Leakage

$$q_{LC} \text{ (gpm)} = (P_I - P_O)(\text{psi}) \times k_L \text{ (gpm/psi)} \qquad 9.3A$$

$$q_{LC} \text{ (lit/min)} = (P_I - P_O)(\text{bar}) \times k_L \text{ [(lit/mi)/bar]} \qquad 9.3B$$

- Leakage coefficient k_L is used to investigate the effect of leakage on the cylinder performance.

- Internal leakage q_{LC} from HP chamber to the LP chamber.
  - Inlet side isn't necessarily the HP chamber

Can you give me an example?

6

6

9.2.1.4- Cylinder Speed

Cylinder **x** and **a** are calculated by integrating/differentiating the speed.

$$v \text{ (ft/s)} = \frac{0.321 \, [Q_I(\text{gpm}) - q_{LC}(\text{gpm})]}{A_I(\text{in}^2)} \qquad 9.4A$$

$$v \text{ (cm/s)} = \frac{1000 \, [Q_I(\text{lit/min}) - q_{LC}(\text{lit/min})]}{60 \times A_I(\text{cm}^2)} \qquad 9.4B$$

9.2.1.5- Cylinder Outlet Flow

For synchronous cylinders, if $q_{LC} = 0$, where $Q_o = Q_I$.

$$Q_O(\text{gpm}) = \frac{v \left(\frac{\text{ft}}{\text{s}}\right) \times A_O(\text{in}^2)}{0.321} + q_{LC} \qquad 9.5A$$

$$Q_O(\text{lit/min}) = \frac{60 \times v(\text{cm/s}) \times A_O(\text{cm}^2)}{1000} + q_{LC} \qquad 9.5B$$

7

7

202

9.2.1.6- Cylinder Force

- **Based on the Source:** *Internal or External.*
- **Based on the Sign:** *Positive (Resistive) or Negative (Assistive).* **Based on the Cylinder Motion:** *Passive or Active.*

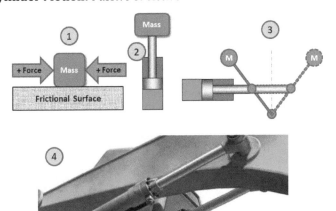

Fig. 9.3- Various Loading Conditions of Hydraulic Motors   8

8

$$F_V(N) = v\ (cm/s) \times k_V \left(\frac{N}{cm/s}\right) \qquad 9.6$$

$$F_I(N) = M\ (kg) \times \dot{v}\ (cm/s^2) \times 0.01 \qquad 9.7$$

$$F_F(N.m) = F_N\ (N) \times k_F(-) \qquad 9.8$$

$$F_C = F_L + F_V + F_I + F_F \qquad 9.9$$

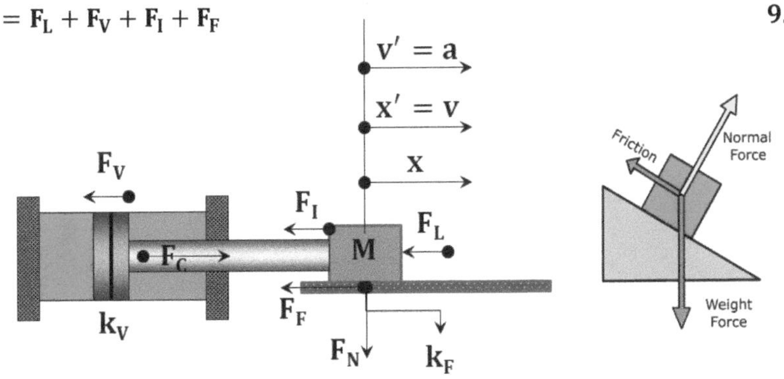

Fig. 9.4- Hydraulic Cylinder Force Calculation   9

9

9.2.2- Simulation Model

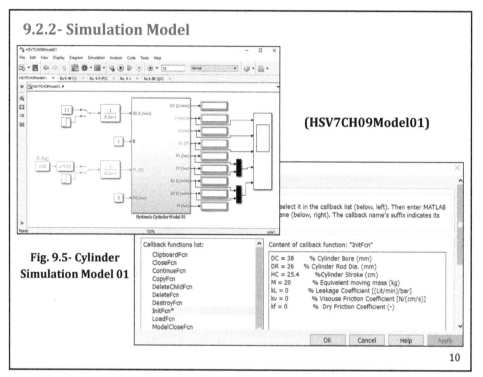

(HSV7CH09Model01)

Fig. 9.5- Cylinder Simulation Model 01

10

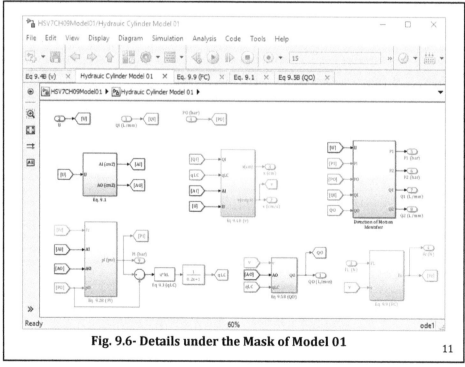

Fig. 9.6- Details under the Mask of Model 01

11

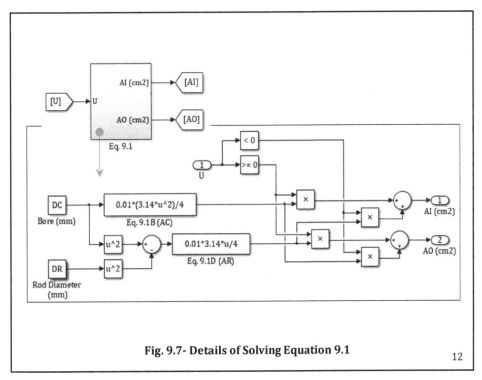

Fig. 9.7- Details of Solving Equation 9.1

12

12

Modeling the Dynamic Effect of Cylinder Load:

- Cylinder load $\rightarrow$ **q$_{LC}$** and **ΔP**.
- Such changes aren't occurred instantaneously.
- 1st order NTF with a reasonable time constant (200 ms.).

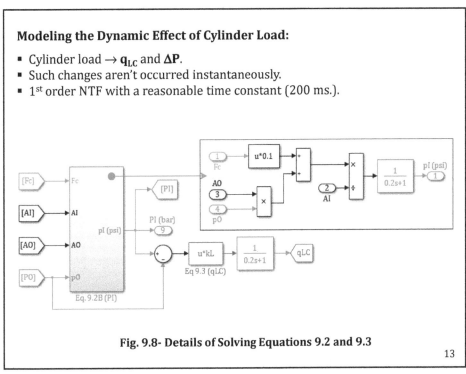

Fig. 9.8- Details of Solving Equations 9.2 and 9.3

13

13

Modeling the Dynamic Effect of Inlet Flow:

- Inlet Flow → **cylinder speed v**.
- fluid compressibility + inertia → **cylinder** isn't moved instantaneously.
- Dynamics developed experimentally.
- The hydraulic cylinder is extended suddenly & dynamics captured.
- Cylinder performs like a first-order system 6.3 inches/s (16 cm/s) in 1 s.
- Time constant equal 1/5 = 0.2 s = 200 ms.

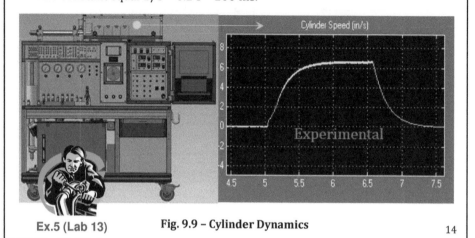

Ex.5 (Lab 13)   **Fig. 9.9 – Cylinder Dynamics**   14

14

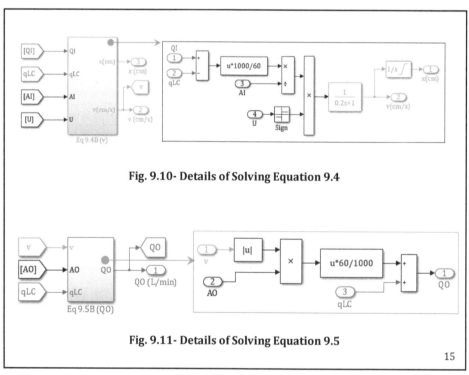

Fig. 9.10- Details of Solving Equation 9.4

Fig. 9.11- Details of Solving Equation 9.5

15

15

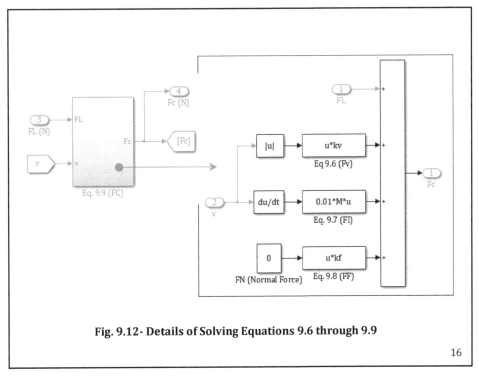

Fig. 9.12- Details of Solving Equations 9.6 through 9.9

16

16

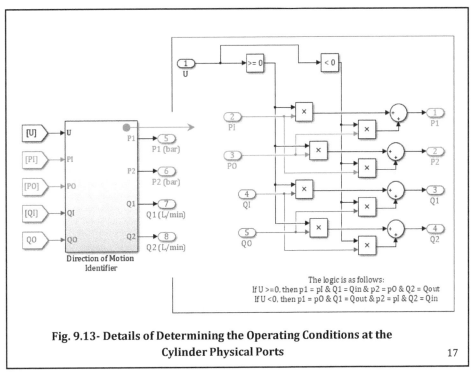

The logic is as follows:
If U >=0, then p1 = pI & Q1 = Qin & p2 = pO & Q2 = Qout
If U <0, then p1 = pO & Q1 = Qout & p2 = pI & Q2 = Qin

Fig. 9.13- Details of Determining the Operating Conditions at the Cylinder Physical Ports

17

17

9.2.3- Model Validation
Steady State Validation:

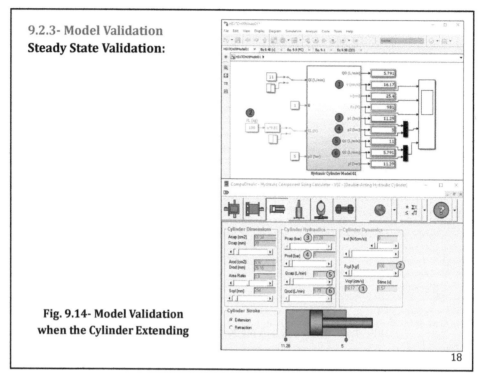

Fig. 9.14- Model Validation when the Cylinder Extending

18

18

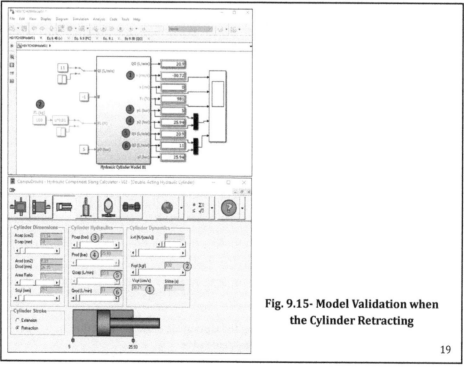

Fig. 9.15- Model Validation when the Cylinder Retracting

19

19

Validating the Dynamic Effect of Inlet Flow:

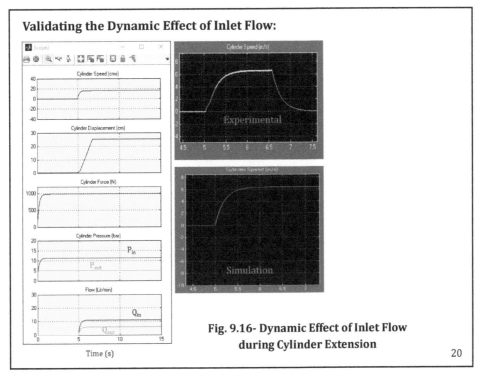

Fig. 9.16- Dynamic Effect of Inlet Flow
during Cylinder Extension

20

20

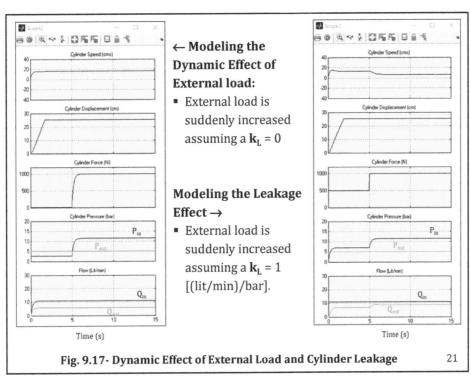

← **Modeling the Dynamic Effect of External load:**
- External load is suddenly increased assuming a $k_L = 0$

Modeling the Leakage Effect →
- External load is suddenly increased assuming a $k_L = 1$ [(lit/min)/bar].

Fig. 9.17- Dynamic Effect of External Load and Cylinder Leakage

21

21

9.3- Modeling Cylinder Drifting due to Oil Bulk Modulus

$$\Delta v = \frac{V_0 \Delta P}{\beta} = \text{Cylinder Area (A)} \times \text{Cylinder Drifting (L)}$$

$$\rightarrow \text{Cylinder Drifting (L)} = \frac{V_0 \Delta P}{A \beta} = \frac{H \Delta p}{\beta} = \frac{HF}{A \beta} \qquad 9.10$$

$$\rightarrow L \text{ (mm)} = \frac{10 \times H \text{ (cm)} \, \Delta P \text{(bar)}}{\beta \text{ (bar)}} \qquad 9.10A$$

$$\rightarrow L \text{ (in)} = \frac{H \text{ (in)} \times \Delta P \text{(psi)}}{\beta \text{ (psi)}} \qquad 9.10B$$

22

22

Example:
- Incompressible fluid $\rightarrow \beta$ = infinity $\rightarrow$ Cylinder Drift = 0.
- A case when Bulk Modulus = 20,000 bar $\rightarrow$
- When P increase 1,000 bar $\rightarrow$ cylinder drifts 5% of the cylinder stroke.

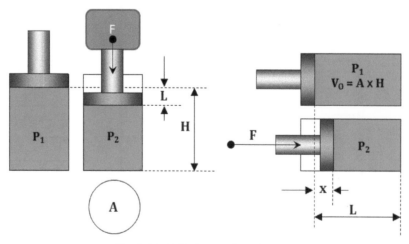

Fig. 9.18- Cylinder Drift Modeling

23

23

9.4- Modeling of Pressure Increase due to Thermal Expansion

Definition of Fluid Thermal Expansion: If a confined volume of liquid is exposed to an increase in temperature, the liquid is thermally expanded intensifying the pressure.

Mathematical Expression:

$$K_{THE} \left[\frac{1}{^{0}C} \right] = \frac{\Delta V/V_0}{\Delta T} \rightarrow \Delta V = K_{THE} \, V_0 \, \Delta T \qquad 9.11$$

$$\Delta v = \frac{V_0 \Delta p}{\beta} = K_{THE} \, V_0 \, \Delta T \rightarrow \Delta P = \beta \times K_{THE} \times \Delta T \qquad 9.12$$

Hydraulic Fluid	Mineral Oil	Phosphate Ester	Water Glycol
Coefficient of Thermal Expansion K_{THE} 1/°C	0.0005	0.00041	0.00034

Table 9.1- Coefficient of Thermal Expansion of Hydraulic Fluids   24

24

Example: A case when Bulk Modulus = 20,000 bar, an increase of temperature of 50 °C in a cylinder filled with mineral oil causes an increase of pressure as follows:

$$\Delta P = \beta \times K_{THE} \times \Delta T = 20,000 \times 0.0005 \times 50 = 500 \, bar$$

- This example shows that pressure can increase to a limit that can damage the cylinder because of oil thermal expansion.

- That is why built in thermal pressure relief valves are used to limit the pressure inside the cylinder.

25

Chapter 9 Reviews

1. When the force acts on the cylinder rod during the cylinder motion and even when the cylinder stops, based on the description in this textbook, the force is named as?
   A. Passive force.
   B. Resistive force
   C. Active force
   D. Assistive force

2. When the force opposes the cylinder motion in both directions, based on the description in this textbook, the force is named as?
   A. Passive force.
   B. Resistive force
   C. Active force
   D. Assistive force

3. When the force pulls the cylinder rod out of the cylinder barrel, based on the description in this textbook, the force is named as?
   A. Passive force
   B. Resistive force
   C. Active force
   D. Assistive force

4. Internal leakage in a hydraulic cylinder resulting in?
   A. Cylinder overheating
   B. Reduced productivity
   C. More energy consumption
   D. All of the above

5. When a cylinder lifting a mass in a vertical position, the load is considered?
   A. Active and Assistive
   B. Active and Resistive
   C. Passive and Assistive
   D. Active and Resistive

Chapter 9 Assignment

Student Name: --- Student ID: ------------------

Date: --- Score: -----------------------

Assignment: Using model (HSV7CH09Model01), investigate the effect of cylinder leakage on the cylinder performance. Use leakage factor $k_L = 2$ (lit/min)/bar

Chapter 10
Modeling of Hydraulic Valves

Objectives:

This chapter presents the lumped modeling concept as applied for hydraulic valves. This chapter considers modeling at least one pressure control valve, one flow control valve, and one directional control valves. Models for electro-hydraulic proportional and servo valves are also developed.

0

0

Brief Contents:

1

1

10.1- Introduction to Hydraulic Valve Modeling

A controller is actuated manually or remotely
(hydraulically by a pilot pressure or EH by an electrical variable signal).

- Modeled based on internal structure for pump development.
- Modeled based reported data for overall machine modeling.

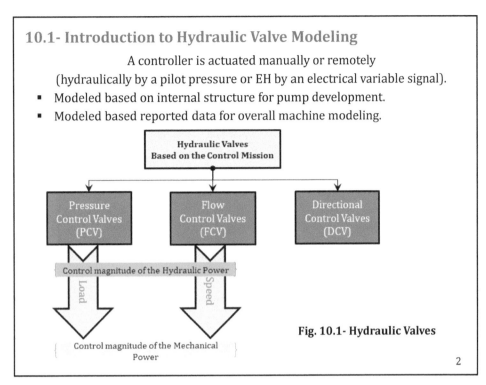

Fig. 10.1- Hydraulic Valves

2

2

10.2- Lumped Model #01 for Pressure Relief Valve Based on Linear Characteristics
10.2.1- Mathematical Model

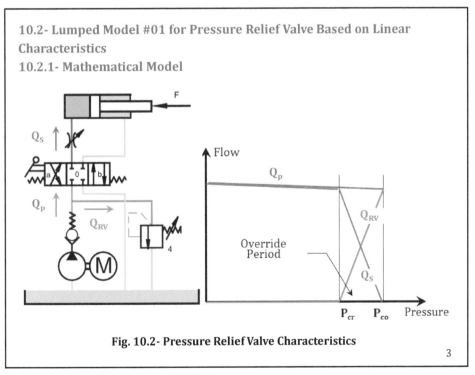

Fig. 10.2- Pressure Relief Valve Characteristics

3

3

$$P_{CO} = P_{CR} + p_{CR} \times \frac{P_{OR}}{100} = p_{CR}\left(1 + \frac{P_{OR}}{100}\right) \qquad 10.1$$

$$\text{If } P_p \leq P_{CR} \rightarrow Q_{RV} = 0 \qquad 10.2$$

$$\text{If } P_{CR} < P_p < P_{CO} \rightarrow Q_{RV} = Q_p \times \frac{P_p - P_{CR}}{P_{CO} - P_{CR}} \qquad 10.3$$

$$\text{If } P_p \geq P_{CO} \rightarrow Q_{RV} = Q_p \qquad 10.4$$

$$Q_S = Q_p - Q_{RV} \qquad 10.5$$

4

4

10.2.2- Simulation Model   (HSV7CH10Model01)

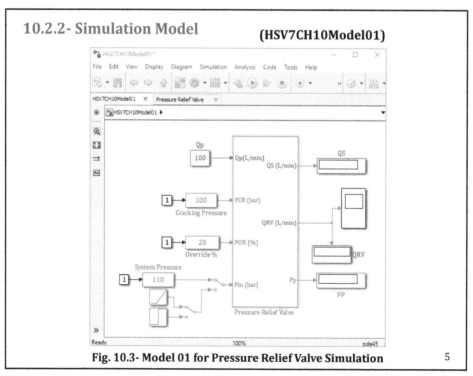

Fig. 10.3- Model 01 for Pressure Relief Valve Simulation   5

5

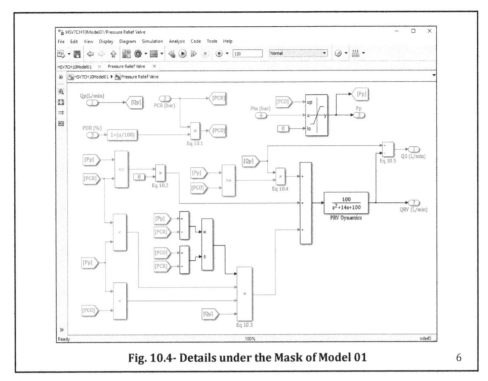

Fig. 10.4- Details under the Mask of Model 01

6

6

10.2.3- Model Validation

Modeling the Valve Steady State "Ramping Pressure"

Modeling the Valve Dynamics "Step Change in Pressure"

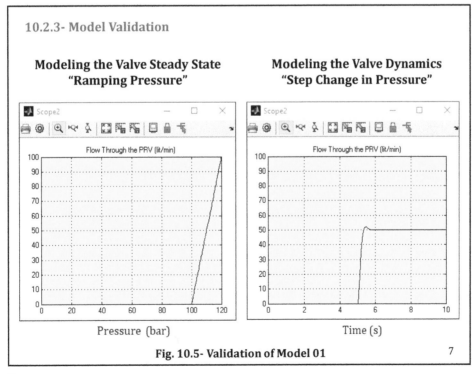

Fig. 10.5- Validation of Model 01

7

7

10.3- Lumped Model #02 for Pressure Relief Vale Based on Nonlinear Characteristics

(HSV7CH10Model02)

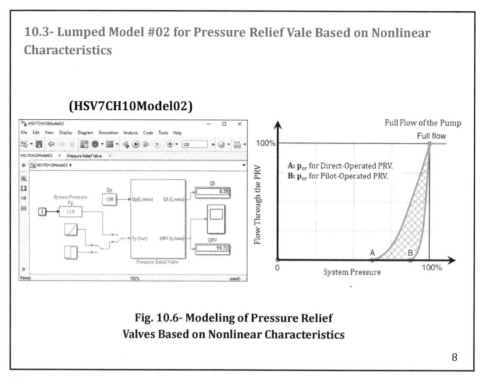

**Fig. 10.6- Modeling of Pressure Relief
Valves Based on Nonlinear Characteristics**

8

8

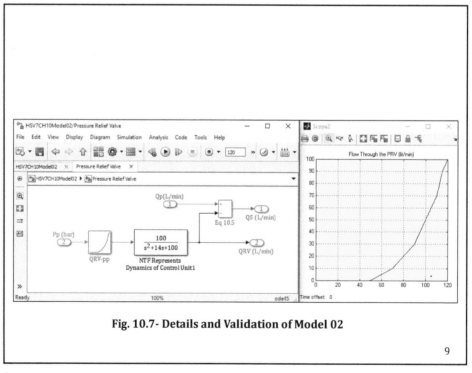

Fig. 10.7- Details and Validation of Model 02

9

9

10.4- Lumped Model #03 for Flow Control Valves
10.4.1- Model Structure and Assumptions

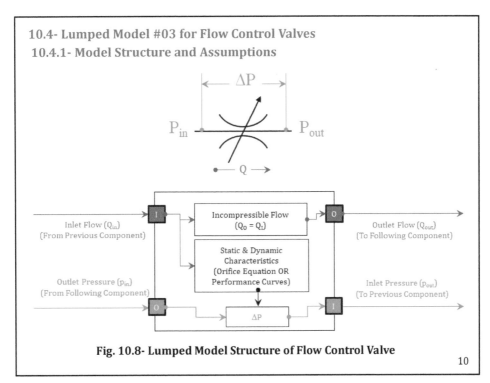

Fig. 10.8- Lumped Model Structure of Flow Control Valve

10

10

10.4.2- Mathematical and Simulation Model   (HSV7CH10Model03)

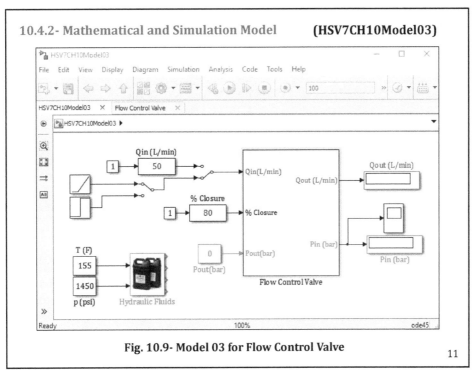

Fig. 10.9- Model 03 for Flow Control Valve

11

11

- saturation block → limit the minimum D_0 → avoid singularity.
- Orifice equation → model is good for all orifice sizes.

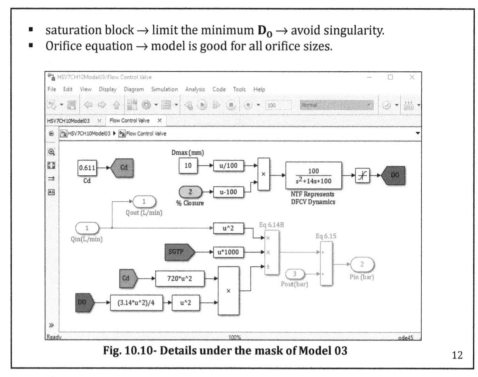

Fig. 10.10- Details under the mask of Model 03

12

12

10.4.3- Model Validation

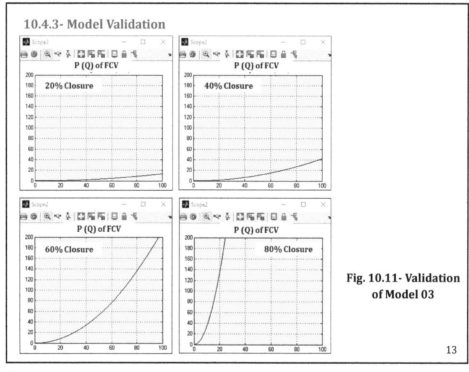

Fig. 10.11- Validation of Model 03

13

13

10.5- Lumped Model #04 for Check Valves   (HSV7CH10Model04)
10.5.1- Mathematical and Simulation Model

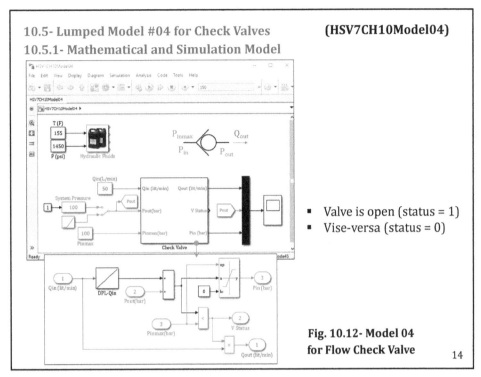

- Valve is open (status = 1)
- Vise-versa (status = 0)

Fig. 10.12- Model 04 for Flow Check Valve

14

14

10.5.2- Model Validation

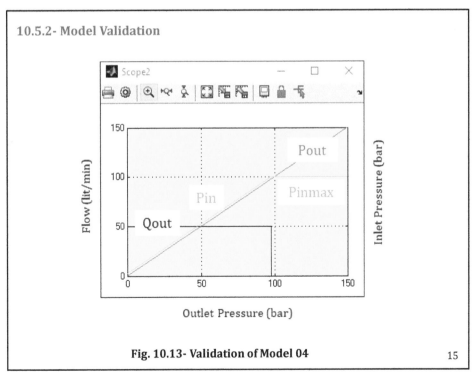

Fig. 10.13- Validation of Model 04

15

15

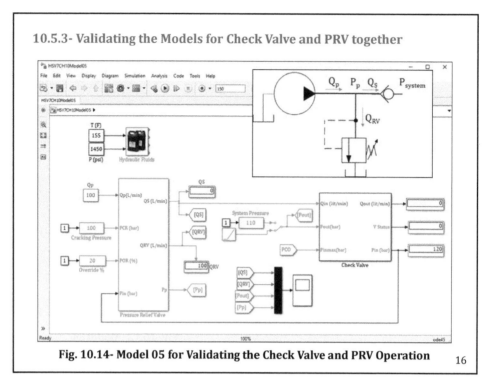

Fig. 10.14- Model 05 for Validating the Check Valve and PRV Operation

16

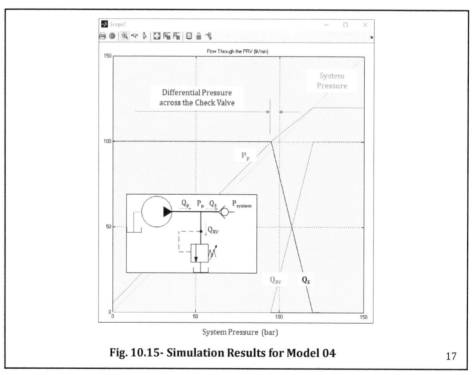

Fig. 10.15- Simulation Results for Model 04

17

10.6- Lumped Model #06 for Continuous Directional Control Valves
10.6.1- Model Structure and Assumptions

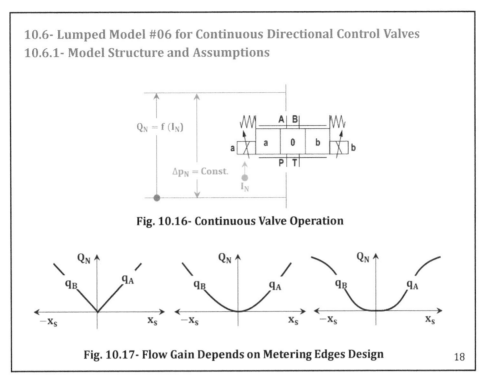

Fig. 10.16- Continuous Valve Operation

Fig. 10.17- Flow Gain Depends on Metering Edges Design   18

18

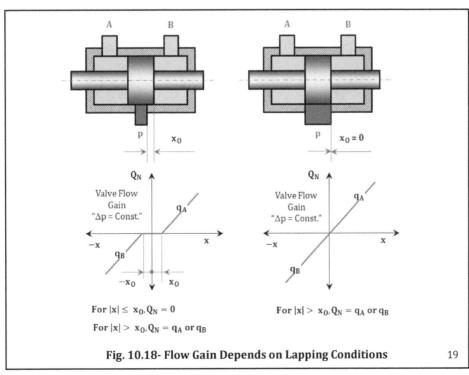

For $|x| \leq x_0, Q_N = 0$

For $|x| > x_0, Q_N = q_A$ or q_B

For $|x| > x_0, Q_N = q_A$ or q_B

Fig. 10.18- Flow Gain Depends on Lapping Conditions   19

19

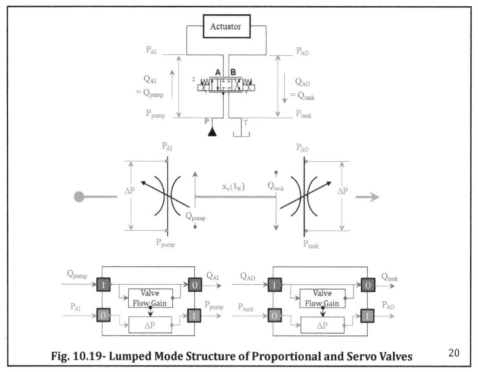

Fig. 10.19- Lumped Mode Structure of Proportional and Servo Valves 20

20

10.6.2- Developing Static Characteristics (Flow Gain) of a Continuous Valve

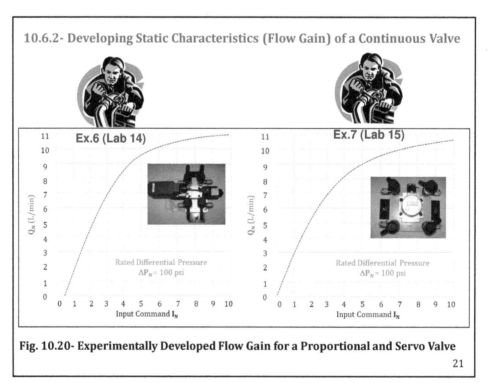

Fig. 10.20- Experimentally Developed Flow Gain for a Proportional and Servo Valve

21

21

10.6.3- Developing Dynamic Characteristics of a Continuous Valve

- 4/3 symmetric valves → identical dynamics in both directions.
- Otherwise, separate dynamic characteristics for each side of the spool movement must be developed.
- Spools valves are spring-centered and surrounded by a damping oil film.
- → So, dynamics of these valves are represented by 2nd order NTF.

By reviewing the data sheets of the valves, it was found that:
- Proportional Valve:
- Natural Frequency = 40 Hz (251rad/s) and damping ratio = 1.

- Servo Valve:
- Natural Frequency = 110 Hz (690 rad/s) and damping ratio = 0.2.

22

22

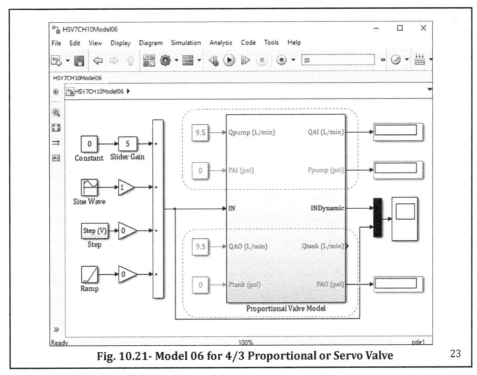

Fig. 10.21- Model 06 for 4/3 Proportional or Servo Valve 23

23

Sequence of calculation:

1. **Defining ΔP_N**
2. **Modeling the Dynamic Characteristics**
3. **Determining Q_N assuming $\Delta P = \Delta P_N$**
4. **Determining Actual ΔP**
5. **Determining P_{pump} and P_{AO} (Eq. 10.8)**
6. **Saturation Blocks:** to avoid unrealistic overshoot of the pressure and flow. For example, during the first iteration, the model receives surge flow that may result is a very large differential pressure. This is just a virtual numerical value that won't happen in reality.

$$\text{Since } Q \propto I \sqrt{\Delta P} \;\rightarrow\; \frac{Q}{Q_N} = \sqrt{\frac{\Delta P}{\Delta P_N}} \rightarrow \Delta P = \Delta P_N \left[\frac{Q}{Q_N}\right]^2 \qquad 10.7$$

$$P_{pump} = P_{AI} + \Delta P_N \left[\frac{Q_{AI} = Q_{pump}}{Q_N}\right]^2 \qquad 10.8A$$

$$P_{AO} = P_{tank} + \Delta P_N \left[\frac{Q_{AO} = Q_{tank}}{Q_N}\right]^2 \qquad 10.8B$$

24

24

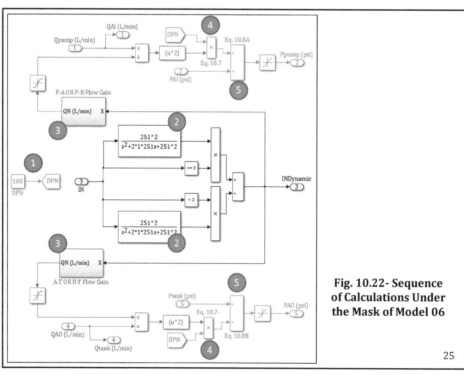

Fig. 10.22- Sequence of Calculations Under the Mask of Model 06

25

25

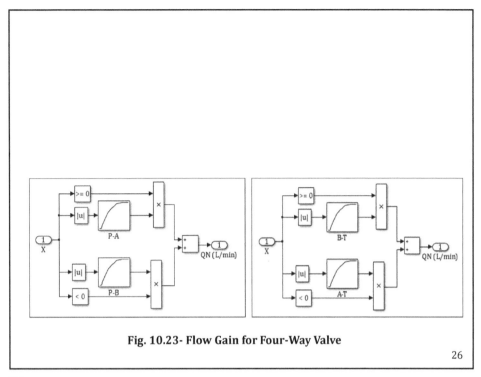

Fig. 10.23- Flow Gain for Four-Way Valve

26

26

10.6.5- Model Validation

- **Point-by-Point Flow Gain Validation**
- Synchronous cylinder
- $\rightarrow Q_{AO} = Q_{AI}$.
- $P_{AI} = 0$ and $P_{tank} = 0$ and
- $Q = Q_N = 9.5$ at $I = I_N$, $= 5$
- $\rightarrow \Delta P = \Delta P_N = 100$ psi.

Fig. 10.24- Point-by-Point Flow Gain Validation

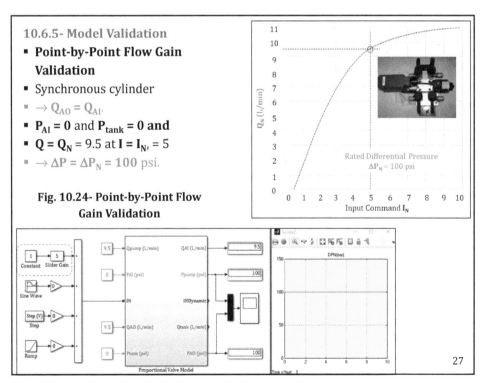

27

27

Flow Gain Curve Validation:

- I_N is ramped from 0-10.
- Simultaneously, the corresponding Q_N is generated from the relevant flow gain and used as input flow.
- $\rightarrow$ constant $\Delta P_N = 100$ psi over the range of I_N.

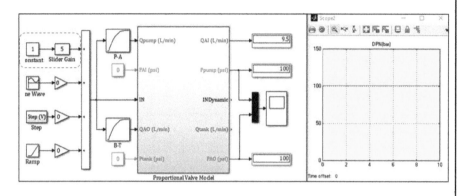

Fig. 10.25- Flow Gain Curve Validation

28

28

Step Response Validation:

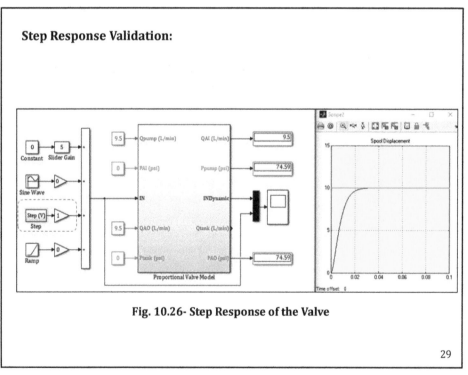

Fig. 10.26- Step Response of the Valve

29

29

Frequency Response Validation:

- When exciting frequency = the natural frequency (40 Hz = 251 rad/s)
- → phase shift = 90 degrees and amplitude ratio = -70% = -3dB.

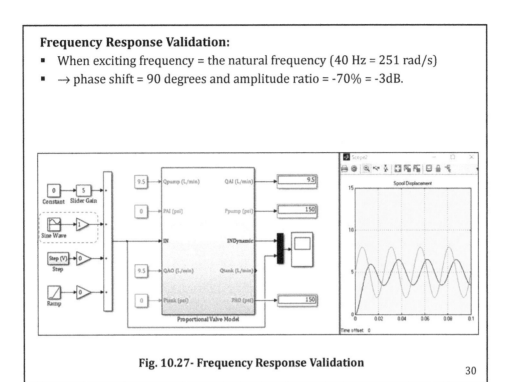

Fig. 10.27- Frequency Response Validation

30

30

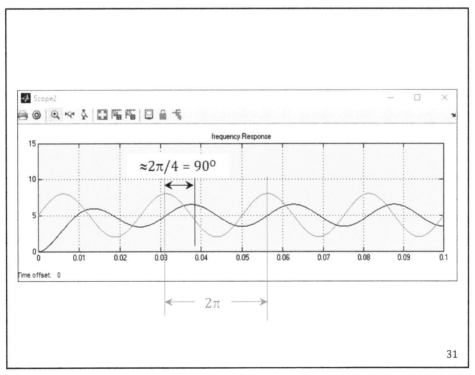

31

31

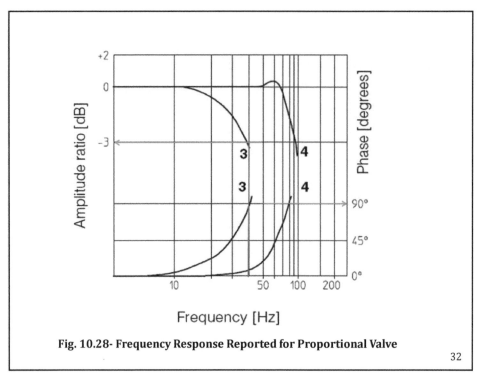

Fig. 10.28- Frequency Response Reported for Proportional Valve

32

32

Chapter 10 Reviews

1. Cracking pressure of a pressure relief valve is the pressure at which?
   A. Valve is fully opened passing all pump flow to the tank
   B. Valve is partially opened passing portion of pump flow to the tank
   C. Valve is critically closed and about to open for further pressure increase
   D. Valve is fully closed passing no flow to the tank

2. A pressure relief valve is critically closed and about to open for further pressure increase?
   A. Pump pressure < PRV cracking pressure.
   B. Pump pressure = cracking pressure
   C. Cracking pressure < pump pressure < cut off pressure
   D. Pump pressure = cut off pressure

3. Flow through a directional control valve depends on?
   A. How widely the valve is opened
   B. Differential pressure across the valve
   C. Fluid specific gravity
   D. All of the above

4. An overlapped spool valve is known as?
   A. A closed center valve
   B. An open center valve
   C. Acritical center valve
   D. All of the above

5. Which of the following proportional valves has the fastest speed of response?
   A. A valve with Bandwidth = 10 Hz
   B. A valve with Bandwidth = 15 Hz
   C. A valve with Bandwidth = 20 Hz
   D. A valve with Bandwidth = 25 Hz

Chapter 10 Assignment

Student Name: --- Student ID: ------------------

Date: --- Score: -----------------------

Assignment 1: Figure 1 and 2 show the static and dynamic characteristics of a proportional directional valve; respectively. Explain, in steps, procedure of modeling the static and dynamic characteristics of the valve.

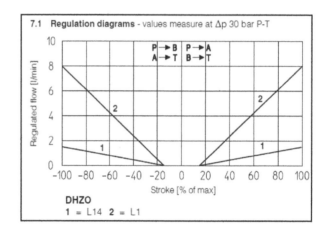

Fig. 1 – Static Characteristics of a Proportional Directional Valve (Courtesy of Atos)

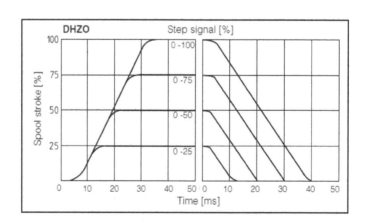

Fig. 2 – Step Characteristics of a Proportional Directional Valve (Courtesy of Atos)

Assignment 2: Using model (HSV7CH10Model06), find the amplitude ratio and phase lag of the valve for an exciting frequency = 500 rad/s

Chapter 11
Modeling of Hydraulic Control Systems

Objectives:

This chapter utilizes the lumped component models previously built to build system models. In this chapter, there is no additional math models to be developed. This chapter assembles system models from component models. After validating system models, they can be used as reference models for purposes of system design or investigating effects of operating conditions on system performance. This chapter presents models for electrohydraulic cylinder position control, electrohydraulic motor speed control, and hydraulic loading system.

0

0

Brief Contents:

11.1-Modeling Electro-Hydraulic Cylinder Position Control System

11.2-Modeling Energy Saving Loading System

11.3-Modeling Electro-Hydraulic Motor Speed Control System

1

1

11.1-Modeling Electro-Hydraulic Cylinder Position Control System

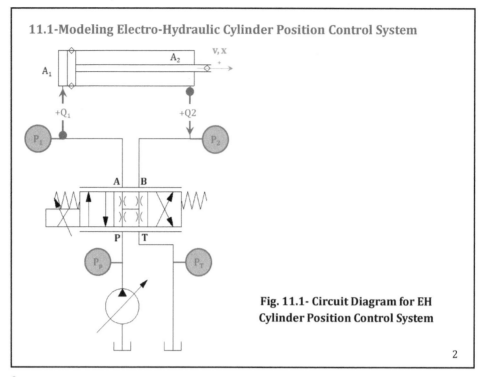

Fig. 11.1- Circuit Diagram for EH Cylinder Position Control System

2

2

11.1.1- Simulation Model

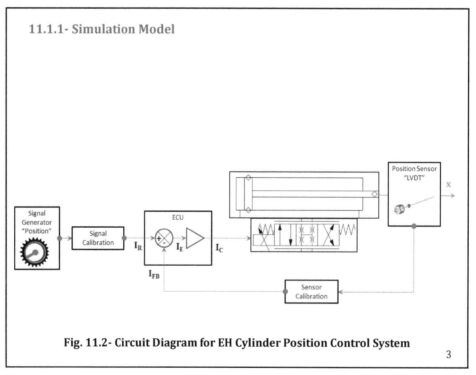

Fig. 11.2- Circuit Diagram for EH Cylinder Position Control System

3

3

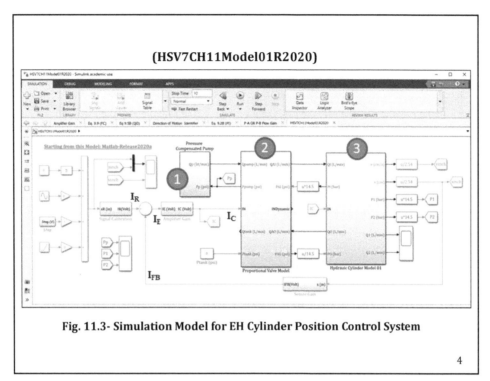

Fig. 11.3- Simulation Model for EH Cylinder Position Control System

4

11.1.2- System Performance Simulation
11.1.2.1- Step Response

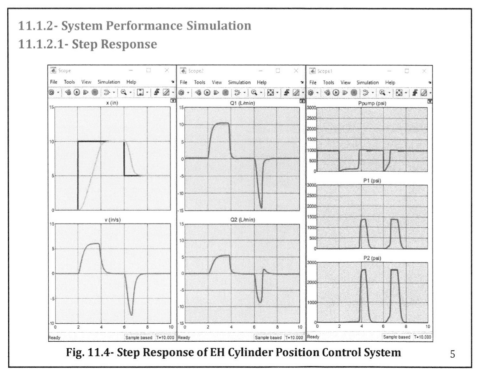

Fig. 11.4- Step Response of EH Cylinder Position Control System

5

11.1.2.2- Harmonic Response

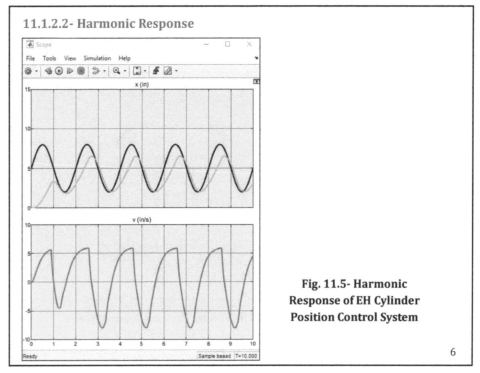

Fig. 11.5- Harmonic Response of EH Cylinder Position Control System

6

11.1.3- Model Validation

Ex.8 (Lab 16)

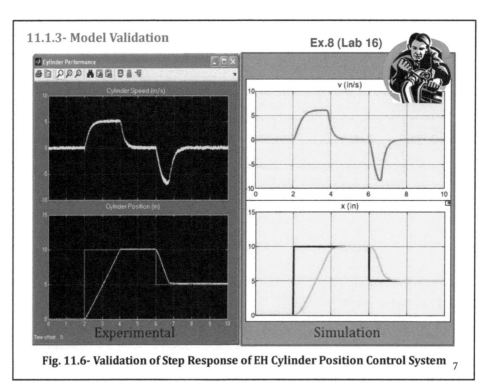

Fig. 11.6- Validation of Step Response of EH Cylinder Position Control System 7

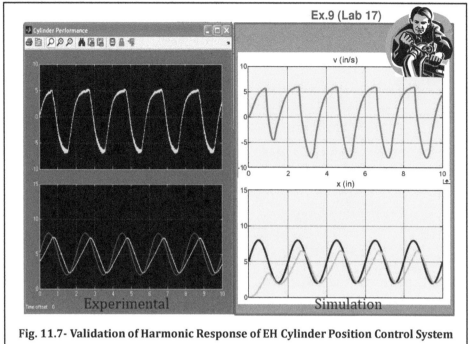

Fig. 11.7- Validation of Harmonic Response of EH Cylinder Position Control System

8

8

11.1.4- Effect of Cylinder Leakage

leakage factor of 2 [(lit/min)/bar] = 0.14 [(lit/min)/psi]

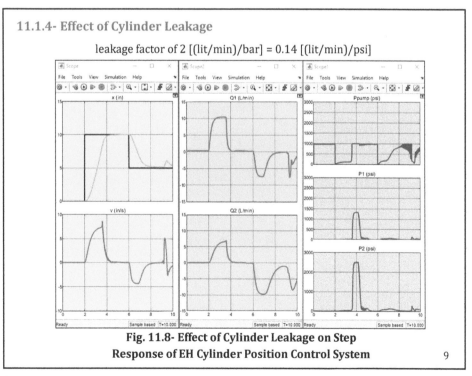

Fig. 11.8- Effect of Cylinder Leakage on Step
Response of EH Cylinder Position Control System

9

9

11.1.5- Effect of Proportional Gain

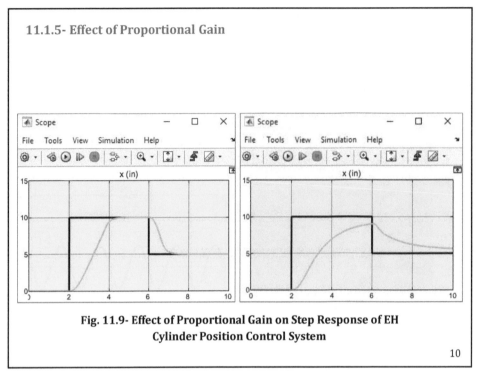

Fig. 11.9- Effect of Proportional Gain on Step Response of EH Cylinder Position Control System

10

10

11.2-Modeling Energy Saving Loading System

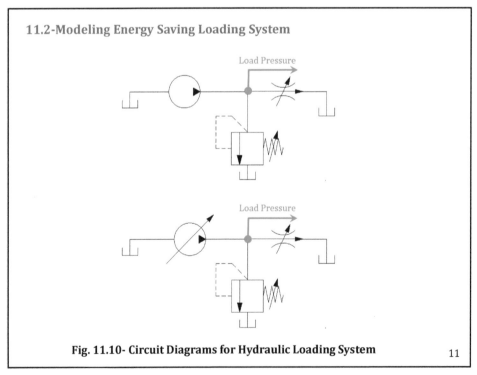

Fig. 11.10- Circuit Diagrams for Hydraulic Loading System

11

11

(HSV7CH11Model02R2020)

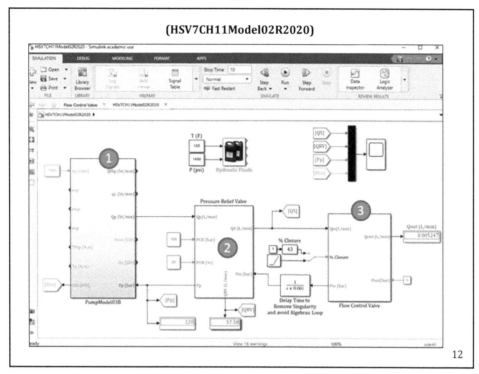

12

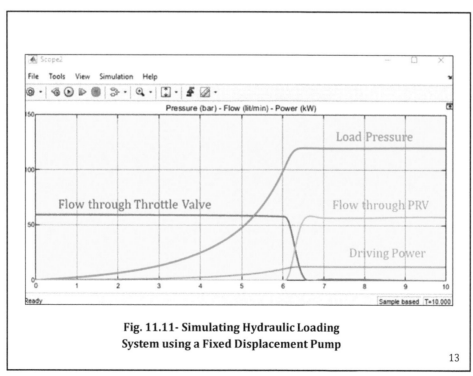

**Fig. 11.11- Simulating Hydraulic Loading
System using a Fixed Displacement Pump**

13

(HSV7CH11Model03R2020)

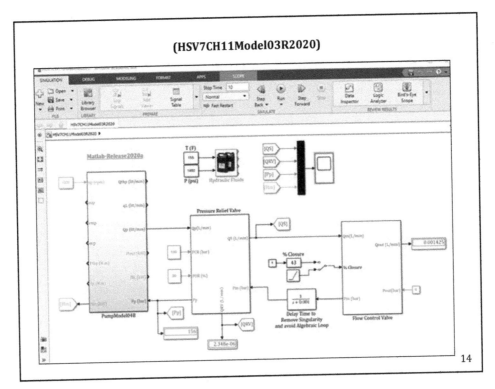

14

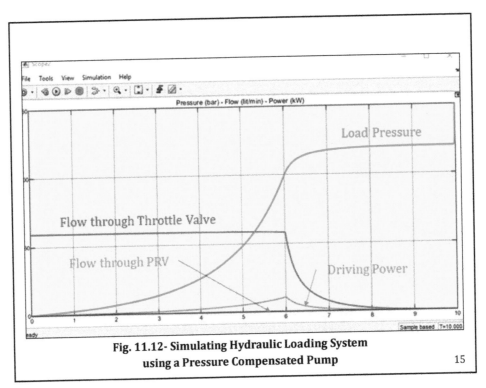

**Fig. 11.12- Simulating Hydraulic Loading System
using a Pressure Compensated Pump**

15

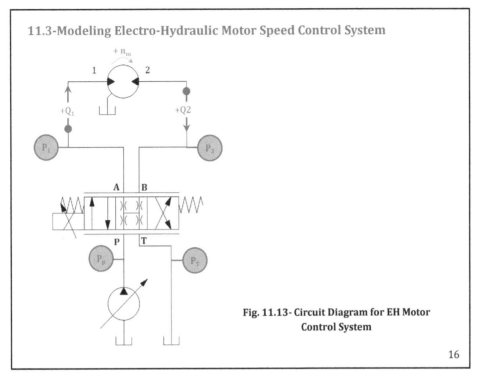

Fig. 11.13- Circuit Diagram for EH Motor
Control System

16

16

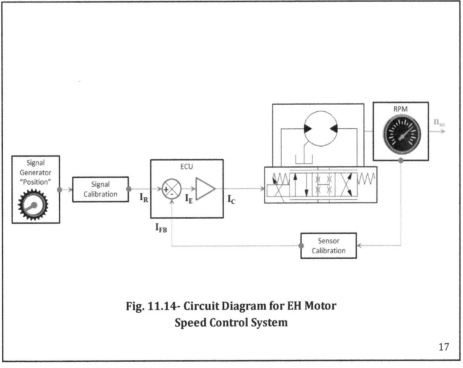

**Fig. 11.14- Circuit Diagram for EH Motor
Speed Control System**

17

17

(HSV7CH11Model04R2020)

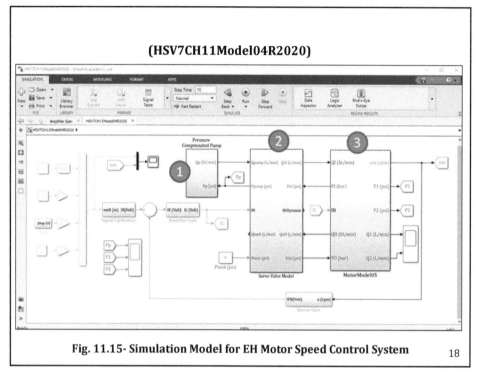

Fig. 11.15- Simulation Model for EH Motor Speed Control System　18

18

11.3.2- System Performance Simulation

11.3.2.1- Step Response

Notes: Every time the motor reverse its motion, control valve passes by its zero position, momentarily the pump is de-stroked/stroked resulting in pressure spikes.

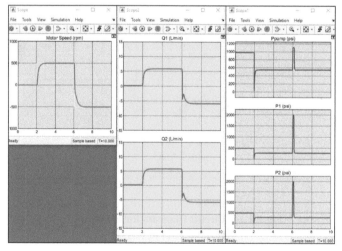

Fig. 11.16- Step Response of EH Motor Speed Control System　19

19

11.3.2.2- Harmonic Response

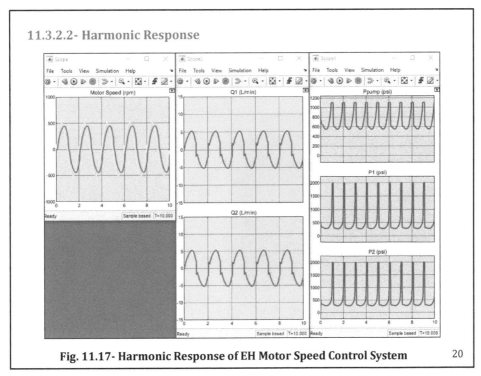

Fig. 11.17- Harmonic Response of EH Motor Speed Control System   20

20

11.3.3- Model Validation   Ex.10 (Lab 18)

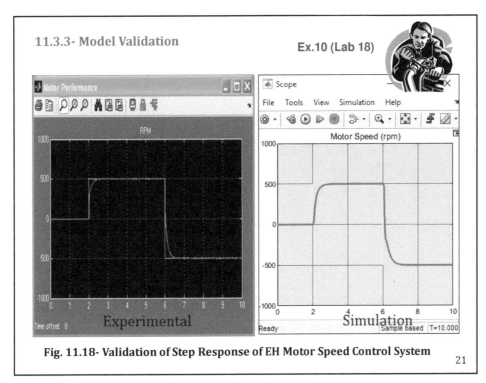

Fig. 11.18- Validation of Step Response of EH Motor Speed Control System
21

21

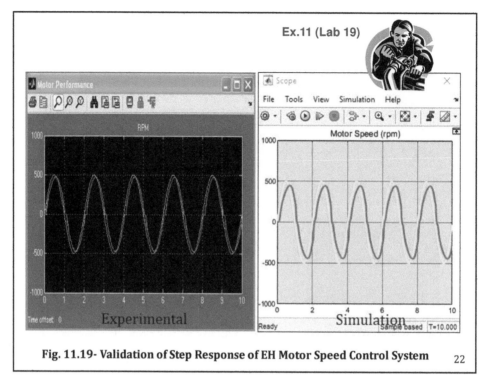

Fig. 11.19- Validation of Step Response of EH Motor Speed Control System

22

11.3.4- Using the Model for System Design

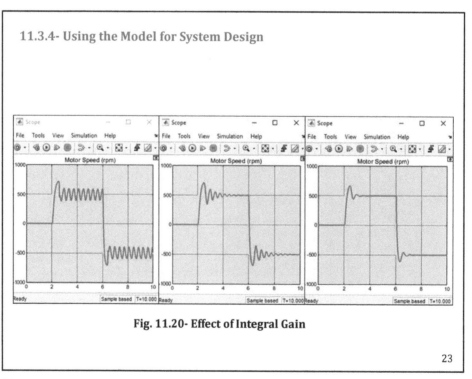

Fig. 11.20- Effect of Integral Gain

23

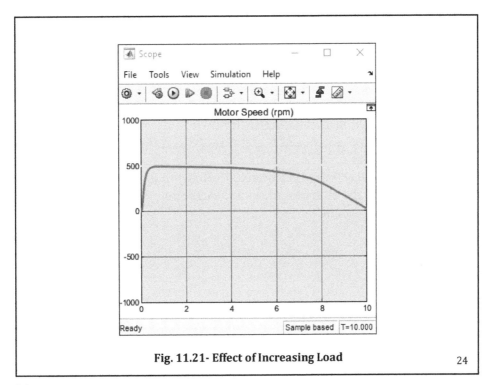

Fig. 11.21- Effect of Increasing Load

24

24

Chapter 11 Assignment

Student Name: --- Student ID: ------------------

Date: -- Score: ------------------------

Assignment: Using model (HSV7CH11Model01R2020) to investigate the effect of the cylinder rod diameter on the step response of the cylinder. Show the effect on the cylinder dynamics and flow requirements

Answers to Chapters Reviews

Chapter 1:

1	2	3	4	5	6	7	8	9	10
D	A	C	A	B	C	D	B	C	A

Chapter 2:

1	2	3	4	5	6	7	8	9	10
D	D	B	C	C	A	B	C	D	A

Chapter 3:

1	2	3	4	5	6	7	8	9	10
B	A	B	D	A	C	B	B	B	B

Chapter 4:

1	2	3	4	5					
D	A	D	B	A					

Chapter 5:

1	2	3	4	5	6	7	8	9	10
D	C	B	B						

Chapter 6:

1	2	3	4	5	6	7	8	9	10
A	B	C	D	A					

Chapter 7:

1	2	3	4	5	6	7	8	9	10
B	C	B	A	D					

Chapter 8:

1	2	3	4	5	6	7	8	9	10
C	A	D	B	C					

Chapter 9:

1	2	3	4	5	6	7	8	9	10
C	B	D	D	B					

Chapter 10:

1	2	3	4	5	6	7	8	9	10
C	B	D	A	D					

Chapter 11:

1	2	3	4	5	6	7	8	9	10

9 780997 763447